plant
A guide to botanical nomenclature
names

Roger Spencer, Rob Cross & Peter Lumley

Third Edition

CSIRO
PUBLISHING

Royal
Botanic
Gardens
Melbourne

www.cabi.org

First edition published 1990
Second edition published 1991
Reprinted with corrections 1995
Third edition published 2007

National Library of Australia Cataloguing-in-Publication entry

Spencer, Roger.
Plant names: a guide to botanical nomenclature.
3rd ed.
Bibliography.
Includes index.
ISBN 9780643094406 (pbk.).
1. Botany – Nomenclature. I. Cross, Robert.
II. Lumley, P. F. III. Title.

581.014

Published exclusively in Australia and New Zealand, and non-exclusively in other territories of the world (excluding Europe, North America, the Middle East, Asia and Africa), by:

CSIRO PUBLISHING
150 Oxford Street (PO Box 1139)
Collingwood VIC 3066
Australia
Tel: 03 9662 7666   Int: +61 3 9662 7666
Fax: 03 9662 7555   Int: +61 3 9662 7555
Local call: 1300 788 000 (Australia only)
Email: publishing.sales@csiro.au
Website: www.publish.csiro.au

Published exclusively in Europe, North America, the Middle East, Asia and Africa and non-exclusively in other territories of the world (excluding Australia and New Zealand), by CABI (CABI is a trading name of CAB International), with ISBN 978 1 84593 374 6.

CABI Head Office
Wallingford, Oxfordshire
OX10 8DE
United Kingdom
Tel: 01491 832 111   Int: +44 1491 832 111
Fax: 01491 829 292   Int: +44 1491 829 292
Email: publishing@cabi.org
Website: www.cabi.org

Cover image courtesy Royal Botanic Gardens Melbourne
Set in 10.5/13 Adobe Goudy and Univers
Cover and text design by Ranya Langenfelds
Typeset by Desktop Concepts Pty Ltd, Melbourne
Printed in China by Bookbuilders

Acknowledgements

Thanks remain to early contributors including Kathy Musial and the late Dr Lawrie Johnson, and especially to the original senior author Peter Lumley for his continued valuable input.

Also thanks to the staff of the Royal Botanic Gardens Melbourne, particularly Professor Jim Ross, Frank Udovicic, Helen Cohn, Neville Walsh and Val Stajsic for critical comments, and to Jill Thurlow for assitance in sourcing images.

Thanks to Jeff Strachan, Plant Variety Protection Office, US; Dr John Wiersema, Curator of GRIN Taxonomy, United States Department of Agriculture/Agricultural Research Service; Dr Arthur Tucker, Delaware State University; Susyn Andrews; Alan Leslie, Royal Horticultural Society, UK; Simon Maughan of the Royal Horticultural Society, UK for permission to reproduce the cover of the The International Clematis Register and Checklist 2002; Helen Costa-Eddy of the Plant Breeders Rights division of Intellectural Property (IP) Australia; Graham Brown of IP Australia (Trademarks Office). The nursery industry experience of Michael Cole, Plant Growers Australia, has been invaluable in developing guidelines for printing names on commercial nursery labels, and we thank him also for supplying sample labels.

The opinions expressed in this book are those of the authors and do not necessarily reflect those of the people mentioned above.

Contents

PART 3 – USING PLANT NAMES 87

Part 4 – Plant Name Resources   iii

Foreword

Most people are introduced to botany and horticulture through plant names. This is important because knowing the correct name for a plant is the key to finding out everything about it.

Unfortunately the use of Latin for botanical names, together with its associated rules and procedures, can seem excessively academic and discourage people from developing a greater appreciation of the world of plants.

Becoming familiar with plant names and understanding the principles underlying their use is an excellent way to make the world of plants more inviting.

Here are some of the most frequently asked questions about plant names:
- Is there anything wrong with using common names?
- Why are botanical names in Latin?
- Who controls their origin and use?
- Why do they change?
- What exactly are cultivars and hybrids?
- Is there a correct way to write and pronounce them?
- How can I remember them all?
- Where can accurate and up-to-date lists of plant names be found?
- Which names and which plants are covered by which *Code* of plant nomenclature?

This booklet will help you with all of these questions.

The introduction discusses wild and cultivated plants and how these have been categorised and named under two plant naming systems or *Codes*.

In Part 1, we examine the use of common names and see how Latin, the original common language of scholars, became established with the development of printing as the international language for plant names. We also see how, later, it became necessary to formulate a set of rules that would ensure consistency in the way names were established and used.

Part 2 explores the difficulties that arose over naming plants that were specially bred or selected for cultivation, and how a similar set of rules became necessary for these plants.

In Part 3, we consider various practical aspects of plant names that are of particular interest to students of botany and horticulture, writers, journalists, plant label manufacturers and others who use botanical names constantly; that is, the way to write, pronounce and remember them.

Part 4 is a resource guide to plant names pointing to further literature and indicating useful websites and places where you can find extensive plant lists and databases.

Foreword to the third edition

This third edition is a response to reader demand for an up-to-date account of plant nomenclature. Since the publication of the first edition of *Plant Names* in 1990 (Lumley and Spencer 1990) and the second in 1991 (Lumley and Spencer 1991), there have been three further editions of the *Botanical Code* (Greuter *et al.* 1994, 2000; McNeill et al. 2006), and two new editions of the *Cultivated Plant Code* (Trehane *et al.* 1995; Brickell *et al.* 2004).

We have included a new introduction to discuss the relationship between the two *Codes* of plant nomenclature and in Part 2 we introduce the idea of the cultigen.

In the last decade there has been a dramatic change in the kinds of names that are printed on nursery labels. Increasingly sophisticated marketing, together with the more widespread use of Plant Breeder's Rights and branding with trademarks, has resulted in a shift of emphasis from botanical and common names to legally protected marketing names, and this has introduced a new set of problems. Consequently, we have extended the section on trade designations (commercial names, many of which are legally protected). Many different kinds of names now appear on nursery labels and so we have made recommendations for presenting these names in a way that distinguishes each different kind of name.

Keeping the names of garden plants stable and encouraging the accumulation of historical information on the origins of cultivars is extremely valuable and so we have included a section on Nomenclatural Standards.

Resources on the internet have improved vastly in the last 5–10 years and this too has been addressed.

All these developments have encouraged us to expand a little on the relevant sections of the book, adding new sections where appropriate.

Roger Spencer, Rob Cross and Peter Lumley
Royal Botanic Gardens Melbourne

Introduction to the Codes of plant nomenclature

If we are to communicate effectively about plants through books, journals, nursery catalogues, databases and general conversation, then we need a precise, stable and internationally accepted naming system.

Plants are named according to the rules and recommendations that are set out in two *Codes* of nomenclature (Figure 1): the *International Code of Botanical Nomenclature* (abbreviated to *Botanical Code*) and the *International Code of Nomenclature for Cultivated Plants* (abbreviated to *Cultivated Plant Code*). Both are formal technical documents that are not easy to read. Nevertheless, they are important because they provide the framework necessary to keep order in the potentially chaotic world of plant names.

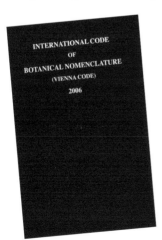

Figure 1: The two *Codes* of plant nomenclature
Image: Rob Cross

I

Why do we have two Codes?

The *Cultivated Plant Code* arose out of the *Botanical Code* about 50 years ago primarily because of the practical need to have simple and stable non-Latin names for those plants of special commercial or ornamental interest that did not fit neatly into the classification categories of the *Botanical Code*.

The two *Codes* also serve the special requirements of different groups of people. The *Botanical Code* focuses on the scientific needs of classification botanists (taxonomists) as they attempt to maintain order and stability for all plant names. Within this overall enterprise the *Cultivated Plant Code* provides for the world of plant commerce: for horticulturists, foresters and agriculturists who deal with ornamental garden plants, timber trees and food crops.

wild
Part one
plants

Common names

Names act as a highly effective shorthand for the objects around us, especially those items that we use regularly or regard as important. Try explaining to someone what happened in a room of 30 people without using the names of the people!

Hunter–gatherers know the plants on which they depend for food, medicine, clothing and tools, but with settled agriculture the world has become progressively urbanised. We are distanced physically and psychologically from the natural environment so that our experience of plant names may be poorer than it has been for generations. Most people know the names of only a few trees, common garden and food plants, and some weeds. A wide-ranging knowledge of plants is very unusual, perhaps only found in some professional horticulturists, keen gardeners, naturalists and botanists. In contrast, many of us are familiar with a range of technical terms used for the parts and functioning of our computers, cars and televisions, simply because we are so directly dependent on them.

Structure

Plant common names have a similar form in most cultures. They are generally composed of one or two words that reflect some aspect of the plant such as its appearance, origin or use. We often name and group objects including ourselves using a noun–adjective binomial, which is a name consisting of two words, one being the name of an object, the other a short description of that object. So, we speak of classes of objects like rice, roses, wattles and

Pythons, and within each class particular individuals might be named; for example, Basmati rice, standard rose, Golden Wattle and Monty Python.

Origin

We assume that most common names were not imposed on people but arose when the need for a name occurred, and they were then maintained through common usage via the direct experience of the plants in nature or gardens, or by word of mouth. Nowadays, we tend to look up the common names of plants in books, except when they are familiar and widely grown garden or food plants.

In Australia, we have adopted many common names that are used in other countries, especially Britain and the United States. These common names used for introduced plants, names like elm, oak, pine and rose, originated long ago in Europe or Asia. The names of some Australian plants, such as Mulga, Wilga, Gungurru and Bangalow Palm (Figure 2) are taken from local Aboriginal languages. Others have names given by the early settlers and refer to their striking appearance, for example Kangaroo Paw and Grass-tree, or they were named for their similarity to European cultivated plants like Native Fuchsia and Willow Myrtle. Trees were sometimes given the names of other trees with similar timber, such as Silky Oak and Mountain Ash.

Common names are still being introduced. One exciting new development in Australia is the acceptance of Asian herb, fruit and vegetable names (often as English translations or transliterations) into our

Figure 2:
Archontophoenix cunninghamiana,
Bangalow Palm
Image: Rob Cross

common name repertoire; for example, Vietnamese Hotmint, Pak Choi, and Star Fruit.

Common names as an alternative to botanical names

For many practical people the Latin system of naming plants appears archaic. Latin is a complicated, unfamiliar and dead language. Latin names also seem to have little relevance to commercial realities. Are they really necessary in the context of, say, a retail nursery? After all, they can be even more confronting to customers than they are to nursery workers, and that does not help sales.

For these reasons it is sometimes suggested that we abandon the unfamiliar Latin and instead use the much simpler common names. In principle, this sounds like a good idea but on closer inspection there are several problems:

- Often there are many different common names for the same plant, and the same name may be used for different plants. Perhaps the commonest of common English names is Lily, which is part of the common name of well over 200 different kinds of plants, and this is followed by names like pea, bean, grass and palm.
- The common name favoured for a particular plant may change over time.
- Most importantly, although we might think we have a grasp of common name usage it is difficult to monitor precisely where and how much a particular common name is being used: common names differ, not only between countries, but also within a particular country, and even from one local area and community to another.
- When a single species is split into two new ones, should both still retain the common name? If 'yes', then how do we distinguish them by the common name? If 'no', then do we invent a new common name? And what happens when *Baeckea behrii*, Broom Baeckea, is transferred to the genus *Babingtonia*?

Botanical names are a way of grouping botanically related plants into families, genera, species and so on. Common names may also be used in this way so we have, for example, brassicas, eucalypts and Thunberg's gardenia, which are the common name equivalents for plants in the botanical categories Brassicaceae, *Eucalyptus* and *Gardenia thunbergii*. However, common names may classify plants in all sorts of non-botanical ways, so they may just as easily give a false impression of plant relationships. The Australian Native Honeysuckle (*Eremophila alternifolia* or *Lambertia multiflora*

THE BIGGEST TREE IN THE BRITISH EMPIRE

or, sometimes, *Banksia*), She-oak (*Casuarina*), and Native Fuchsia (*Eremophila* or *Correa*), for instance, are botanically unrelated to their exotic namesakes Honeysuckle (*Lonicera*), Oak (*Quercus*) and *Fuchsia*. The common name Mint is based on a plant's smell and flavour, and therefore does not always apply to plants in the genus *Mentha*, the culinary mint genus. We categorise food plants into vegetables and fruits, and garden plants by their garden function as a windbreak, groundcover or climber. Eggs-and-bacon is a name given to almost any Australian native plant with red and yellow pea-like flowers; although all these plants are in the botanical family Fabaceae, it is the red and yellow colouring that is an equally important factor in determining the common name.

The following example illustrates the difficulties associated with using the popular common name Mountain Ash. The Mountain Ash of the Australian state of Victoria, *Eucalyptus regnans* (Figure 3), is so called because its timber resembles that of the European Ash, *Fraxinus excelsior*. In Tasmania, it is known as the Swamp Gum, a name that in Victoria is generally given to *Eucalyptus ovata*. In England, the Mountain Ash is a small upland tree with ash-like leaves and red berries, *Sorbus aucuparia*, which in Scotland is called Rowan. In America, the Mountain Ash is *Sorbus americana*. You see the problem!

The *Cultivated Plant Code* deals with the hotch-potch of different kinds of common names by distinguishing between: *colloquial names*, those used in local communities but not widely enough to be recorded; *common names*, the non-scientific names widely used and recorded in a particular area; and *vernacular names*, those translated from scientific names into the local language.

Latin botanical names overcome all this confusion because there is only one botanical name for each kind of plant, even though that name might change from time to time! The principle of one name for one kind of plant is universally appealing and important regardless of whether the names we are using are commercial, legal or scientific. With modern marketing, a nursery worker will insist that nobody uses his or her legally protected names or company trademarks, and that proper databases and records be kept to

Figure 3: (Left) Mountain Ash, *Eucalyptus regnans*, in Victoria, Australia, one of the world's tallest trees
Image from: Ray C (1932–1933). *The World of Wonder*. Amalgamated Press, London.

ensure that people can distinguish his or her goods from those of others. Botany has been trying to do this for the entire Plant Kingdom for well over 250 years.

Attempts have been made to avoid Latin by developing a 'one plant one name' approach to common names, an approach that may simplify databasing. This is done either by inventing names, or attempting to regulate them by producing standardised lists in which only one common name is provided for each species (or a preferred common name is suggested). For example, English translations of the Latin names are sometimes used. *Mentha rotundifolia* might be translated and listed as Round-leaved Mint even though this name may never have been in common usage. This is a way of giving common names to the many plants in Australia that do not already have them. Although this avoids the problem of using Latin, it creates other difficulties. Who chooses the preferred name to be adopted in cases like this, why is a particular name preferred and how is everyone to find out the 'accepted' common name?

Of course, botanical names may be used as common names: *Azalea* was once the botanical name for what we now know as a section of *Rhododendron*, and we use the botanical words chrysanthemum, camellia and fuchsia in the same way as though they were common names.

Historical and cultural value

Common names may be romantic (Love-in-a-mist, Angels' Tears, Forget-me-not, Love-lies-bleeding) or down-to-earth (Chicken Gizzard, Bastard Balm, Giant Hogweed, Stinking Roger). They are a simple, often charming or evocative, way of referring to plants. Also, they frequently have historical, cultural or other associations that would be a pity to lose. For these reasons they have greater general appeal than the apparently difficult botanical names. There is no doubt that they will always be used.

Nothing is wrong with common names except their lack of precision. The botanical name is the only one that clearly identifies a particular kind of plant, and that can be understood across language and regional barriers.

Figure 4 has extracts of several verses from Iris Bayley's *West Indian Weed Song* demonstrating the wide use of common names in the West Indies.

Figure 4: (Right) Iris Bayley's *West Indian Weed Song*

West Indian Weed Song

Iris Bayley

(Quoted in Julia F. Morton (1981) Atlas of Medicinal Plants of
Middle America. Charles C. Thomas, Springfield, Illinois, U.S.A.)

One day I met a old woman selling, and I wanted something to eat
I thought I could put a little bit in she way, but I take back when I did meet.
I thought she had bananas, orange or a pear, nothing that I need,
I asked the old woman what she was selling, she said she was selling weed.

She had de Cassava-mama, Okra-babba, Jacob-ladder, mixed with Finegona,
Job-tea, Peter-parslee, John-Belly-parslee and the White Clary,
Bill-bush, Wild-cane, Duck-weed, Aniseed, War-bitters and Wild-grey-root,
She even had down to a certain bush Barbados call Puss-in-Boots.

She had de Pap-bush, Elder-bush, de Black-pepper bush, French-to-you
and de Cure-for-all,
Sapodilla, Tamarind-leaf, Money-bush, and de Soldier-parsley,
Pumpkin-blossom, with Double-do-me, and Congo-pumps in galore,
Physic-nut, and even the Lily-root is the list of her everyday soup.

The Pitons, Martinique

Illustration from Villiers-Stuart (1891) Adventures amidst the equatorial forests
and rivers of South America; also in the West Indies and the wilds of Florida to
which is added 'Jamaica revisited'. John Murray, London.

Latin names, the binomial system and plant classification

When the first printed books were circulated, Latin was the internationally accepted language of Western scholars because herbalists and botanists relied on the classical Greek texts and their Latin translations for plant descriptions. Using Latin names allowed communication about plants across the different European languages. Once Latin names became established throughout Europe it would have been impractical to revert to using common names. Eventually, Latin became accepted world-wide for plant names (Figure 5).

By the 18th century, Latin names for plants consisted of short descriptions, called diagnoses, which enabled readers to identify and distinguish one plant from another. So in 1738 the great Swedish biologist Linnaeus named the catmint *Nepeta floribus interrupte spicatis pedunculatis*, meaning '*Nepeta* with flowers in a stalked interrupted spike'.

In 1753 Linnaeus produced a definitive list of plants titled *Species Plantarum* (Species of Plants). Like his contemporaries, he used diagnoses, but in this book he put a single index word next to each diagnosis. For catmint it was *cataria* (Figure 6). This index word was combined with the name *Nepeta*, the heading under which a series of diagnoses was written, to form a two-word name referred to as a binomial. From this time on, the use of binomials like *Nepeta cataria* became established, and *Species Plantarum* became the starting point for modern botanical names.

Linnaeus intended to classify, describe and name all living organisms. His 'Sexual System' placed plants into groups based on the number of stamens and styles in the flower. It was a rather 'artificial' system based

Figure 5: An example of the convenient international use of Latin from the *Flora of China*
From: Wu T (Ed.) (1981). *Flora Reipublicae Popularis Sinicae Volume 16(2)*. Science Press, Beijing

on only a few characters but it was easy to use. Another example of a simple artificial system would be the grouping of plants according to their flower colour.

During the 18th and 19th centuries various attempts were made to provide a more 'natural' system; one in which plants with many characters in common were classified together. As the theory of evolution became accepted by biologists in the 19th and 20th centuries, 'natural' classifications used increasing numbers of characters and became more intent on grouping together plants closely related in an explicitly evolutionary sense. As part of the normal scientific process, new classification systems are published from time to time to indicate new views on the relationships between plant groups.

NEPETA.

cataria. 1. NEPETA floribus fpicatis, verticillis fubpedicellatis, foliis petiolatis cordatis dentato-ferratis.
Nepeta floribus interrupte fpicatis pedunculatis. *Hort. cliff.* 310. *Hort. upf.* 163. *Fl. fuec.* 486. *Roy. lugdb.* 316. *Gron. virg.* 65. *Mat. med.* 291. *Dalib. parif.* 116.
Mentha cataria vulgaris & major. *Bauh. pin.* 228.
Cataria herba. *Dod. pempt.* 99.
Habitat in Europa. ♃

Figure 6: The description of *Nepeta cataria* in Linnaeus' *Species Plantarum*, 1753
From: Linnaeus, Carl (1753). *Species plantarum...* Laurentii Salvii, Holmiae.

The *International Code of Botanical Nomenclature*

At the end of the 19th century, the Western world's botanists had reached a common understanding of the structure of plant names and the ranks in the hierarchy used for plant classification. In the 20th century, these ideas were formalised in the *International Code of Botanical Nomenclature*. This first *Botanical Code* was formulated at a Botanical Congress in Vienna in 1905 and it is now reviewed every six years. The *Botanical Code* is administered by the International Association for Plant Taxonomy.

Botanical names in Latin form are now accepted internationally and offer a precise and efficient means of communication. The preamble to the *Botanical Code*, slightly simplified, states:

> Botany requires a precise and simple system of nomenclature used by botanists in all countries, dealing on the one hand with the terms which denote the ranks of plant groups and on the other hand with the scientific names which are applied to the individual plant groups. The purpose of giving a name is not to indicate the character of a plant group but to supply a means of referring to it and to indicate its rank. This code aims at the provision of a stable method of naming plant groups, avoiding and rejecting the use of names which may cause error and ambiguity.

Principles of the Botanical Code

Following the preamble are six principles with a set of articles (rules) and recommendations. These prescriptions are accepted internationally by botanists but they are not legally binding.

Figure 7: *Morus nigra*, the black mulberry

Image from: Stephenson J and Churchill J (1831). *Medical botany.* John Churchill, London.

Figure 8: *Morus serrator*, the Australian gannet

Image from: Gould J (1848). *Birds of Australia Volume 7.* John Gould, London. (Reproduced with permission from the collection of the Library, Museum of Victoria.)

Principle 1 states simply that plant nomenclature, animal nomenclature and bacteriological nomenclature are independent. This means that it is possible to have the same name for two quite different organisms. *Morus* according to the *Botanical Code* is the mulberry genus (Figure 7), while *Morus* according to the *Zoological Code* is the gannet genus (Figure 8).

Principle 2 is important but rarely understood by non-botanists. It prescribes that the names of plants or plant groups are based on TYPES which, with rare exceptions, are actual dried specimens of plants (Figure 9).

Principle 3 states that nomenclature is based on priority of publication. This principle stresses the overriding importance of the first published name and, together with Principle 4, provides a means of determining which of several published names for the same plant is correct.

TYPES

The name used for a particular plant or group is based on one particular specimen, the type specimen (and its assigned replicates) stored in a dried-plant repository called a herbarium. For instance, the TYPE of *Callistemon pearsonii* is a dried, pressed specimen housed in the National Herbarium of Victoria at the Royal Botanic Gardens Melbourne. It remains as a reference against which other specimens are compared. Principle 2 highlights the importance of herbarium collections and of type specimens in particular; it also accounts for the fact that most botanists spend more time with dead, dried specimens than with living plants.

Figure 9: The type specimen of *Callistemon pearsonii* R.D.Spencer & P.F.Lumley housed in the National Herbarium of Victoria
Image: Carl Davies, CSIRO

HOLO-
TYPE

NATIONAL HERBARIUM OF
VICTORIA (MEL), AUSTRALIA

MEL 1535969

MEL1535969

NATIONAL HERBARIUM OF VICTORIA (MEL)
MELBOURNE, AUSTRALIA

Callistemon pearsonii R.D. Spencer &
P.F. Lumley

Family: MYRTACEAE
Loc.: QUEENSLAND - Blackdown Tableland,
Mimosa Creek.

Lat. 23°38'S Long. 149°00'E
Coll.: R.D. Spencer, no. 84
14th October 1984

Det.:
Notes: Growing along Mimosa Creek, mostly
less than 1 m. tall, (occasionally up
to 2 m. tall) with dark, quite deeply
fissured bark. Very attractive red
brushes. Capsules not persisting for
more than a year.

Dupl.: BRI, NSW

AUTHORS AND PUBLICATION

Principle 3 implies that botanical names must be *published*. When a new plant is described and named, the name and description are published in a recognised, printed journal or book. This means that plant names have authors (often called 'authorities'): the people who first validly published the name of the plant. For instance *Nepeta cataria*, named by Linnaeus in *Species Plantarum*, is known as *Nepeta cataria* L. (L. is the conventional international abbreviation for Linnaeus). The abbreviation 'F. Muell.' can be seen after many Australian native plant names because Ferdinand Mueller, Director of Melbourne's Royal Botanic Gardens from 1857 to 1873, described over 2000. Although authors are of little interest to the general user of plant names, they are important in distinguishing between the same name given independently to two different species (homonym). Author names are usually abbreviated using standard abbreviations as listed in *Authors of plant names* (Brummit and Powell 1992).

THE PRINCIPLE OF PRIORITY

The so-called 'Principle' or 'Rule' of Priority referred to in many publications is a blend of Principles 3 and 4. The maidenhair tree is a straightforward example of the Principle of Priority (Figure 10). It was given the name *Ginkgo biloba* by Linnaeus in 1771 (*Ginkgo* being a transliteration of the Japanese form of the Chinese name), but the word *Ginkgo* was considered to be uncouth by Sir James Smith, who coined a new name in 1797: *Salisburia adiantifolia*. This name became widely used in the nursery trade and is seen in old nursery catalogues; however it is incorrect, according to the *Botanical Code*, because Linnaeus' name has priority of publication.

The Principle of Priority causes annoyance because familiar names are replaced when earlier published names for the same plant are discovered; however, this convention is not always followed. Many subsequently published names have been retained by being 'conserved' in a special appendix to the *Botanical Code*, even though they do not have priority. Examples are *Banksia*, *Telopea*, *Grevillea* and the economically important *Triticum aestivum*, better known as wheat.

Figure 10:
Salisburia adiantifolia became the widely used name in the nursery trade for *Ginkgo biloba* after being coined by Sir James Smith in 1797, but the Principle of Priority required *Ginkgo biloba* to be reinstated.

Image from: Veitch J and Sons (1881). *A manual of the Coniferae*. H.M. Pollett & Co., London.

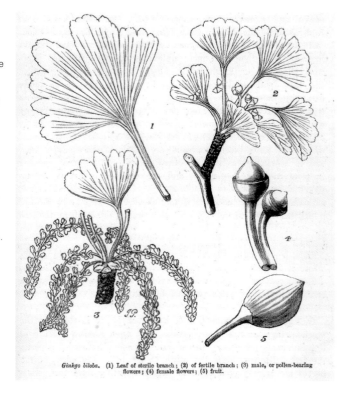

Ginkgo biloba. (1) Leaf of sterile branch; (2) of fertile branch; (3) male, or pollen-bearing flowers; (4) female flowers; (5) fruit.

Principle 4 prescribes that each plant or group of plants within a particular system of classification can bear only one correct name: the earliest one following the rules.

Principle 5 states that scientific names are to be treated as Latin. So, a plant named after Ferdinand Mueller is called, for example, not *Eucalyptus mueller* but *Eucalyptus muelleriana* where the name is given a standard Latin suffix.

Principle 6 makes the rules of the *Botanical Code* retrospective.

The botanical hierarchy

Biological nomenclature attempts to provide a simple way of giving names to organisms, and to do this without making any assumptions about the methods, purposes or principles of taxonomy. It does not concern itself, for example, with the reasons for making particular groupings or the kinds of characters used to distinguish those groupings. The *Botanical Code* does, however, assume that plant groups are arranged in a nested hierarchy, like 'boxes within boxes'.

The nested hierarchy

The Plant Kingdom is organised into groups of plants with similar characteristics and each group is a sub-set of a larger, more inclusive group. So, within a particular classification system each species is included within one, and only one, genus; each genus in one, and only one, family, and so on. The more inclusive the group, the higher up we are in the hierarchy. One useful result of a system of this sort is that it is predictive: in knowing that a plant belongs to a particular group you will also know that it shares many features with the other members.

For the most part, the nested hierarchy system of naming organisms works very well, presumably because it reflects the way organisms have evolved by the modification of existing structures. However, as we shall see, it doesn't easily accommodate hybrids, cultigens and ranks below the level of species.

Ranks and taxa

The preamble to the *Botanical Code* states that it deals 'on the one hand with terms which denote the ranks of plant groups and on the other hand with the scientific names which are applied to the individual plant groups'.

The words 'taxonomic group' are used so frequently that they have been contracted to the word 'taxon' (pl. taxa). Taxa are assigned to a particular level within the classification hierarchy, known as a rank. Myrtaceae, *Eucalyptus*, *Eucalyptus globulus* and *Eucalyptus globulus* subsp. *maidenii* are all examples of taxa and they are assigned to the ranks of family, genus, species and subspecies, respectively (Figure 11).

Historically, although many different classification systems have been proposed, there has been a commonly accepted hierarchy of ranks. This has also occurred in some human organisations, such as the armed forces and business companies, and makes comparison of ranks relatively easy. No doubt the uniformity between different ranking systems has arisen partly through convenience, and partly because of an understanding of the optimum size of groups in a classification.

The most commonly used ranks in botany are the lower ranks of genus and species. Above the genus is the family and below the species are the subspecies and variety. The *Botanical Code* lists 24 ranks, with the lowest being the subform and the highest the kingdom, but very few of these ranks are used regularly, even by botanists. Ranks such as orders and families can be recognised by their word ending so, for example, orders end with *-ales* as in Rosales, and families end in *-aceae* as in Rosaceae, and so on. However, the endings of genus (generic) names and specific epithets do not have identical endings.

RANKS AND TAXA

It can be confusing that we use the words 'species', 'family', etc., to denote both ranks and taxa. *Acacia pycnantha* is a species (using 'species' in the sense of a particular group of plants, a taxon) at the rank of species (using the word 'species' as the name of the rank of the taxon).

Order

An order, which always has the ending *-ales* (meaning 'belonging to'), is a group encompassing a number of families. Orders are rarely referred to in general botanical or horticultural books.

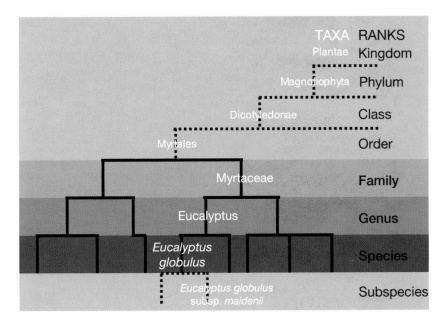

TAXA	RANKS
Plantae	Kingdom
Magnoliophyta	Phylum
Dicotyledonae	Class
Myrtales	Order
Myrtaceae	Family
Eucalyptus	Genus
Eucalyptus globulus	Species
Eucalyptus globulus subsp. maidenii	Subspecies

Figure 11: The nested hierarchy of ranks and taxa. The most commonly referred to ranks have darker background shading.

FAMILY

Closely related genera are grouped together at the rank of family. Taxa at the rank of family are often used in botanical writing and are easily recognised because they all have the ending -*aceae* (meaning 'resemblance'). The *Botanical Code* prescribes that a family name is formed from the name of the type genus by adding -*aceae*, thus *Poa* gives rise to the grass family Poaceae and *Rosa* to Rosaceae. Two widespread and diverse families in Australia are the Myrtaceae (including eucalypts, bottle-brushes, paperbarks and tea-trees) and Proteaceae (including banksias, hakeas, grevilleas and waratahs) (Figure 12). Family names are occasionally used in gardening books.

A few family names may be the source of confusion because there are differing botanical opinions on how families should be organised within orders. Some botanists recognise particular families within their classification system while others do not, or different botanists may recognise a particular family as containing a different suite of genera. These different systems reflect differing views about the relative importance of the various characters that separate one family from another. For most families, however, there is agreement among botanists.

There are currently several similar family classification systems used in the world. These systems assist with presenting plant descriptions in floras and the arrangement of dried specimens in herbaria (see Classification systems, Part 4). The *Flora of Australia* (Flora of Australia Editorial Committee 1981–) and *Flora of North America* (Flora of North America Editorial Committee 1993–) use the system of American botanist Arthur Cronquist (Cronquist 1981), while *Flora Europaea* (Tutin *et al.* 1964a, 1964b, 1968, 1972, 1976, 1980) is largely based the system of Engler-Diels. Additional systems include those of the Royal Botanic Gardens, Kew (Brummitt), Dahlgren (especially his treatment of monocotyledons), Thorne, and others.

Genus

From the earliest times people seem intuitively to have grouped similar plants that correspond to the botanical use of the term 'genus' (plural genera). Probably for that reason it is the most easily comprehended group. The genus consists of one or more kinds of plants that share a distinctive set of characters. Its name is a singular noun in Latin form, such as *Rhododendron*, *Fuchsia*, *Chrysanthemum*, *Lavandula*, *Quercus* and *Eucalyptus*. The different kinds of plants within a genus are called species.

Species

The species is the basic unit of classification. When someone asks for the name of a plant, the answer is usually a species name. As we have seen, the species name is a binomial: the name of the genus followed by a specific epithet; for example, *Nepeta cataria*.

FAMILY NAME ALTERNATIVES

A number of old established family names, which are exceptions to the -aceae family-ending rule, are allowed in the *Botanical Code*. The following are legitimate alternative names for the same family:

Compositae (daisy family)	Asteraceae
Cruciferae (cabbage family)	Brassicaceae
Gramineae (grass family)	Poaceae
Guttiferae (hypericum family)	Clusiaceae
Labiatae (mint family)	Lamiaceae
Palmae (palm family)	Arecaceae
Umbelliferae (carrot family)	Apiaceae

The old family Leguminosae (the legume or pea family) is sometimes split into the families Papilionaceae, Mimosaceae and Caesalpiniaceae. Unfortunately, the name Fabaceae is used by some botanists as an alternative for Papilionaceae and by others as an alternative for Leguminosae. Our recommendation is that the old family Leguminosae be now recognised as the families Fabaceae (peas or beans), Mimosaceae (wattles) and Caesalpiniaceae (cassias) in accordance with the *Flora of Australia*.

The family Liliaceae until recently was taken by many botanists to encompass familiar horticultural families such as the Amaryllidaceae. In recent years, research has resulted in the fragmentation of this broad family into a smaller, narrowly defined Liliaceae and many other families including Amarayllidaceae, Agapanthaceae, Hyacinthaceae, Alstroemeriaceae, Colchicaceae, Asparagaceae, Alliaceae, Convallariaceae.

It is very difficult to define a plant species. At one extreme it may be understood in a practical way as a group of plants that can be distinguished from other species and to which a competent botanist gives a binomial, a practical pigeonhole for the purposes of identification and communication. This is in fact the way that most species have been, and still are, established.

At the other extreme, a theoretical definition would be that members of a species actually or potentially interbreed but do not normally breed with other species. Because of the wide range of breeding behaviour, this definition is not satisfactory for plants, but it illustrates the importance of the species concept in our understanding of plant evolution.

The following three categories are below the level of species (infraspecific) and are given in order of rank. Although the order of ranks is clear, there is no universal agreement about the kinds of plants to be placed in each category. Increasingly, the only infraspecific rank used is the subspecies as with, for example, the *Flora of North America*.

Subspecies

The subspecies is generally understood as having defining characteristics that are usually geographically separated, although they may occupy different ecological niches. For example, *Eucalyptus leucoxylon* subsp. *leucoxylon* is the typical subspecies of south-western Victoria and South Australia, while *E. leucoxylon* subsp. *connata* differs in minor but distinctive features, and is found in the Brisbane Ranges, in populations on the western side of Port Phillip Bay, Victoria, and Studley Park, Melbourne.

WRITING NAMES OF SUBSPECIES

Note that the epithet for the typical subspecies is the same as that for the species. In this case, the author is also automatically the same and is by convention not repeated after the subspecies name. Thus we have:

Eucalyptus leucoxylon F. Muell. subsp. *leucoxylon*

Eucalyptus leucoxylon F. Muell. subsp. *connata* K. Rule

Variety

A variety is often understood as having characters that differ in a minor way from the usual characteristics of the species but plants with these characters do not have a clearly defined geographical or ecological distribution. Thus, *Eucalyptus ovata* var. *grandiflora* is a large-flowered variant of the typical variety *Eucalyptus ovata* var. *ovata* (usually simply written as *Eucalyptus ovata*). A variety may be very common.

The word 'variety' has a definite botanical usage and it is therefore confusing when it is used occasionally in horticulture to refer to any plant, whatever its rank, as when people speak of a plant nursery full of interesting 'varieties'.

Form

This category is now rarely used but was generally applied to botanically trivial differences, such as an occasional variation of flower or foliage colour, and often of sporadic occurrence; for example, *Cedrus atlantica* f. *glauca,* the blue atlas cedar, which has blue foliage.

It is possible for all three of the above infraspecific (below species level) ranks to be used in the same name in rank order, so we have for example *Prunus donarium* subsp. *speciosa* var. *nobilis* f. *sirotae.* In such instances the name is usually reduced, so in this case it would be written *Prunus donarium* f. *sirotae.*

Botanists nowadays rarely use more than one infraspecific rank within a particular species. Where the distinguishing characteristics are confined to a particular geographic area then subspecies is generally used. Where the minor differences are local or ecological, variety is mostly used. This practice obscures the difference in rank between these two categories, as does the current trend among botanists to use subspecies in preference to other infraspecific categories.

Natural hybrids

Hybrids result from the interbreeding of related species, almost always in the same genus. This may occur when, in nature, the distributions of the two species overlap. *Eucalyptus* × *studleyensis* is a natural hybrid between *Eucalyptus camaldulensis* and *E. ovata* that was discovered in Studley Park, in the Melbourne suburb of Kew.

Although many hybrid taxa have been named in the past, it is rare nowadays for known new natural hybrids to be given botanical names; they are generally referred to, and written, simply as hybrids between the two parent species:

Eucalyptus camaldulensis × *E. ovata*

Sometimes plants are named without the author realising that they are natural hybrids. If their hybrid ancestry later becomes clear then the name is usually retained but an '×' is placed in front of the epithet.

(For cultigen hybrids, see page 56.)

Name changes

Changes in plant names are, understandably, seen by many people as frustrating and unnecessary. The worst examples – when a plant swaps its name with another, or when there is a return to a previous name – have earned botanists an unfortunate reputation. Nursery owners are also well aware that people tend to buy plants with well-known names, which is no incentive to keep their names up-to-date.

There are three basic reasons why plant names are changed by botanists: the first is *nomenclatural* (to conform with the rules of the *Botanical Code*); the second is *taxonomic* (as a result of a revised view of plant relationships), and the third is to correct a misidentification or misapplied name.

Nomenclatural changes

There was great confusion in the botanical literature by the middle of the 19th century. Botanists in countries that were sometimes at war with one another were describing and naming the tens of thousands of plants discovered in the burst of European colonial expansion. The same species was given different names; different species were given the same name.

The *Botanical Code* eventually provided a framework for restoring order to this chaos. Nomenclatural changes are made to ensure that the names of genera and species conform to the rules of the *Botanical Code*. A change results in a correct new name (i.e. it obeys the rules of the *Botanical Code*). The formerly used name is then referred to as a 'synonym'. The *Botanical*

Figure 13:
Phragmites australis (syn. *P. communis*).
Image from: Reichenbach HGL (1850). *Icones Florae Germanicae et Helveticae, Volume 1.* 2nd edition. Friedrich Hofmeister, Leipzig.

Code has been a stabilising influence as the number of described flowering plant species has risen to about 250 000.

We have already seen the application of Principles 3 and 4 of the *Botanical Code* to the Maidenhair Tree *Ginkgo biloba* (see page 19). Another example is the Common Reed (Figure 13). This is the familiar tall, grassy, almost bamboo-like plant growing along the banks of rivers and streams; one of the most widespread plants of the world, being found on all non-Antarctic continents and in all states of Australia. Its former botanical name, *Phragmites communis*, would be familiar to most botanists, but an earlier name and description has been found, so the Principle of Priority requires that its name be changed to *Phragmites australis*, under which it was first described. *Phragmites communis* is therefore now a synonym and should not be used.

There are various other ways in which names might require changing to comply with the *Codes*, as when under the *Botanical Code* they are illegitimate or invalidly published, or under the *Cultivated Plant Code* they have not been properly established, and under both *Codes* when the names are incorrectly spelled according to the rules.

Taxonomic changes

Taxonomy is the study of the principles and procedures of classification. A plant taxonomist reconsidering the classification of a particular group of

Figure 14:
Botanists use
herbarium
specimens to
help classify and
describe the Plant
Kingdom
Image: Frank Udovicic

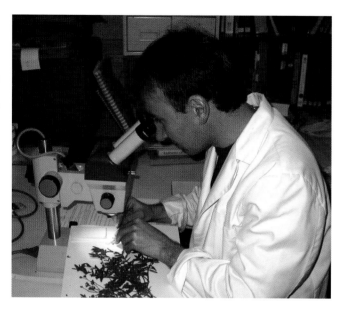

Figure 14:
Botanists use
herbarium
specimens to
help classify and
describe the Plant
Kingdom
Image: Frank Udovicic

plants has certain aims: to clearly delimit the taxa; to discern natural relationships; to produce a practical classification; and to ensure that plant names are correctly applied in relation to the type specimens.

A plant group is often chosen for study because of difficulties in assigning plants to a particular species (identification problems) or because of new investigative techniques that give fresh insights into plant relationships. Examples of the latter include new technologies that increase the range of plant characters that can be studied beyond the traditional morphological studies using a light microscope (Figure 14). Modern techniques include molecular studies such as DNA sequencing, ultrastructural investigations using electron microscopy, chemical analysis of tissues and plant extracts like flavonoids and isozymes, and so on. Modern computer software is used to analyse large data sets of plant characters to produce hypothetical evolutionary trees: this study is referred to as cladistics (*cladus* = branch).

Figure 15: (Right) Herbarium specimens can be stored and curated for a long time with little deterioration, so a large selection of specimens can be assembled for taxonomic study. This specimen of *Banksia serrata* was collected in 1770 by Banks and Solander at Botany Bay.
Image: Royal Botanic Gardens Melbourne

ISO-
LECTOTYPE

MEL 583558

MEL583558

Banksia serrata, L.f. "Saw Banksia"
Ex Herbario Musei Britannici.

NEW HOLLAND
BANKS & SOLANDER
1770
Isostylis serrata Britten
Illustr. Bot. Cook's Voyage, tab.270 p.55
8.23
(*Botany Bay*, April 1770.)

ROYAL BOTANIC GARDENS
AND
NATIONAL HERBARIUM, MELBOURNE
VICTORIA, AUSTRALIA

Banksia serrata L.f.

Loc.: NEW SOUTH WALES. Botany Bay.

Coll.: Banks & Solander, April 1770.

Det.:

Notes:

NATIONAL HERBARIUM OF
VICTORIA (MEL), AUSTRALIA

Figure 16: Herbarium specimens are stored in folders in specially designed archival cupboards in climate- controlled buildings. These specimens in red folders at the Royal Botanic Gardens Melbourne are the type specimens for species in the genus *Acacia*.

Whatever techniques are used, large numbers of herbarium specimens that have been collected from natural habitats in many different localities, and often over a long period of time, are examined in detail to determine the extent of natural variation (Figure 15). If possible, the characteristics of live plants are also studied in nature.

A careful analysis is then made of all the similarities, differences and relationships of the specimens under examination. Once the botanist has determined what range of variability is acceptable for a particular taxon, he or she then selects the type specimen(s) (Figure 16) that fit within this range of variation so that the taxon can be named accordingly.

After all this research, it may well be the opinion of the taxonomist that two species would be better regarded as one, or that one species should be split up into two or more others. It is also quite likely that new species will have been discovered and require description. Usually these studies result in more taxa being formed (more species or genera), but an exception may occur in a group which has been poorly understood in the past. It might be shown that many of the names previously used refer to the same species, and are therefore synonyms.

In this way the group of plants, often a genus, is examined more exhaustively than ever before. The conclusions are then published in a scientific book or journal and are generally known as a revision or monograph of the genus.

DESCRIBING A NEW SPECIES

Some of the principles and rules we have presented here can be illustrated by a further example. In 1986 two of the authors of this book described a new species of bottlebrush from the Blackdown Tableland in south-east Queensland (Figure 17). This 'species' had been known for many years but was undescribed. An attractive plant with crimson brushes and bright yellow anthers, it was available in the nursery trade as *Callistemon* (Blackdown) (see page 59 for this form of citation) or *Callistemon* (Mimosa Creek). After a careful examination of the plant, and a comparison with the types and specimens of other callistemons, a formal description of the plant was written in both Latin and English. This was published in 1986 according to the rules of the *Botanical Code*, together with an illustration, in *Muelleria*, the scientific journal of the National Herbarium of Victoria. Its full citation (name and reference) is: *Callistemon pearsonii* R.D. Spencer & P.F. Lumley: *Muelleria* 6(4): 293–298 (1986)

The specific epithet *pearsonii* refers to Steven Pearson, for many years the Ranger on the Blackdown Tableland who, with his wife Alison, did much to document and photograph the fascinating plants of this area. A type specimen was deposited at the National Herbarium of Victoria (Figure 9), and duplicate specimens of the same plant were sent to herbaria in Brisbane and Sydney. This full citation indicates that this species meets the requirement of Principle 3 for formal publication and also tells us the authors and the place where the description can be found: it is done for accuracy, completeness and ease of verification.

Figure 17: The flower, habit and habitat of *Callistemon pearsonii* at Blackdown Tableland in Queensland. *Callistemon pearsonii* was named after Steven Pearson, a Ranger who for many years documented the flora of the National Park.

Images: Roger Spencer

RECLASSIFICATION, RANKS, NAMES AND THE PRINCIPLE OF PRIORITY – AN EXAMPLE

There is some confusion about the Principle of Priority. Like all parts of the *Botanical Code* it is concerned with name-forming and ranks, not particular systems of classification. Here is a more complex example that illustrates a combination of nomenclatural and taxonomic procedures and principles.

The Brush Box, until 1982 known as *Tristania conferta* R.Br. (R.Br. means Robert Brown), was transferred to a new genus *Lophostemon* by two

Sydney botanists, Wilson and Waterhouse, because in their opinion it differed in sufficiently important ways from other *Tristania* species to be placed in a separate genus. It is now known as *Lophostemon confertus* (R. Br.) Peter C. Wilson and Waterhouse. Note that in the move to the new genus the original specific epithet *conferta* has been kept, together with the author who gave the epithet, Robert Brown, whose name now appears in brackets, followed by the authors of the new name (known as a new combination) *Lophostemon confertus*. It is the original epithet *confertus* that has priority, although the group in which this plant is classified has been changed. Note also that according to the rules of Latin grammar, epithets must agree in gender with the genus they describe and this accounts for the difference in name endings (*confertus* is masculine, to agree with the masculine *Lophostemon*, and *conferta* is feminine, to agree with the feminine *Tristania*).

In accordance with Principle 4 of the *Botanical Code*, only one name is permissible within a particular classification system (see page 20). In *Tristania*, it is the original name *Tristania conferta* and in *Lophostemon* it is *Lophostemon confertus*. Which of the names becomes accepted in the long run is a matter of acceptance by other botanists. In this particular case, there has been general acceptance of the change. Occasionally, there are differences of opinion as in the case of whether or not to use the name *Corymbia* for the bloodwood eucalypts, or how to split up orchid genera. These taxonomic disagreements have nothing to do with the *Botanical Code* as the names at issue have been formed in accordance with its nomenclatural rules.

RESOLVING CONFLICTING TAXONOMIC VIEWS

The natural variation found in plant populations, the gradual divergence and evolution of species, the tendency for hybridisation, and human-induced changes which affect normal breeding patterns all make it difficult to pigeon-hole and describe taxa. In difficult cases, the plant taxonomist must use his or her professional judgement to produce the best objective classification possible based on the evidence at hand. This inevitably leads to some arbitrary decisions or disputed conclusions, which may be changed at a later date. The description of a new species or other taxon may be regarded as a hypothesis that may or may not survive the test of time.

The published description gives other taxonomists an opportunity to assess the new information. In Australia, for example, there has been a heated debate concerning the transfer of about 100 species of *Eucalyptus* to the genus *Corymbia*.

In a similar way, there are a number of different overall classifications of the Plant Kingdom by botanists who interpret current knowledge in different ways. Their views might differ on which plants are to be put into which groups: they might also differ on the particular rank that should be used for particular groups. This can be frustrating for students keen to find a definitive classification system. However, it is hardly surprising that there are divergent opinions about something so complex as the classification of the Plant Kingdom.

There is no international committee or person that makes a final decision in these cases and in this way taxonomic changes differ from nomenclatural ones. A revision will become accepted and used only if it is generally regarded as satisfactory by other botanists who use the changes in their publications. This does not mean that anything is permissible: scientific journals are refereed to make sure that any proposed changes are soundly based.

One last example will help explain another aspect of these changes.

AN EXAMPLE OF A NAME CHANGE FOR TAXONOMIC REASONS

Plants currently known and sold as *Eucalyptus lehmannii* are mostly *E. conferruminata*. This situation has arisen because *E. conferruminata* was only described in 1980 and was previously included in the species *E. lehmannii*. It just happens that most plants available in the trade are those now placed in *E. conferruminata*, not *E. lehmannii*. As this change was not widely publicised, the outdated name has persisted. This kind of taxonomic change poses a particular problem in checking nomenclature. Although we know that most plants in the trade are *E. conferruminata*, some may in fact be correctly named *E. lehmannii* because they are from populations of this species.

Whenever a taxonomic decision of this particular sort is made, whether at family, genus, or species level, the original name still applies to some individuals. The taxonomic decision alters the number of individuals within the new concept of the species, increasing the diversity when species are 'lumped' together or decreasing it when species are 'split' up (Figure 18).

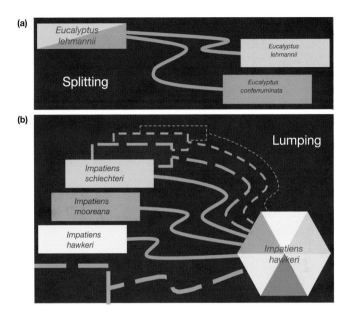

Figure 18: The splitting and lumping of taxa. **a:** The original concept of *E. lehmannii* has been changed by creating a new species, *E. conferruminata*, thus reducing the number of individuals and range of diversity under the name *E. lehmannii*. The type specimen of *E. lehmannii* must, of course, fall within the limits of the new understanding of *E. lehmannii*. **b:** In 1980, 15 groups of New Guinea balsams were recognised, some of which corresponded to published descriptions under names such as *Impatiens hawkeri*, *I. schlechteri*, *I. mooreana* etc. Hybrids between members of these groups are sold as 'New Guinea hybrids'. It was considered that these groups were part of the natural variability within a single species found across the island of New Guinea, and that although these groups may eventually evolve into discrete species, the groups did not yet warrant species status. The name which has priority for this broadly-defined species is *I. hawkeri*, which in botanical literature is indicated as *I. hawkeri* senso lato (s.l.) meaning 'in the wide sense'. When the name *I. hawkeri* was originally published it represented only a proportion of the balsams growing in New Guinea and this sense of the name is represented botanically as *I. hawkeri* sensu stricto (s.s.) meaning 'in the narrow sense'. Under the new concept, the New Guinea hybrids are all infraspecific and could be treated as cultivars of *I. hawkeri* s.l.

Misidentifications and misapplied names

Changes in plant names may sometimes be needed to correct a simple error, as when labels are misplaced in a nursery. But, for the most part, changes become necessary when a plant has been misidentified or its name has been misapplied. The difference between misapplied names and misidentifications is rather subtle but is relevant to both botany and horticulture and requires some explanation.

Plant identification is the act of determining the name of a plant (botanists actually refer to this formal process as 'determination'). Of course, plant identification is not always an elaborate formal process: it also occurs when someone asks for, and is given, the name of a plant in a nursery. Clearly, a full botanical assessment and a quick decision made in a nursery are rather different but, as we generally understand it, a misidentification occurs when, after due consideration, a plant is assigned an incorrect name.

A name is said to be misapplied when it is *used* in a different sense from that intended by the original author of the name. In horticulture, some names are simply garden inventions. For example, the name *Sempervivum nigrum* has no basis under the *Botanical Code* but has been used for a garden plant. Names like this may be listed in reference books as 'a name of no botanical standing'. In another case, a plant might have been known to the gardening community for many years, sometimes decades, under the wrong name, a name that has been perpetuated in published material such as nursery catalogues and gardening books. The evergreen alder, a tree known in Australia for many years under the name *Alnus jorullensis*, is now known by its correct name *Alnus acuminata* var. *glabrata*. The name *Alnus jorullensis* is a true botanical name but the plants in Australia with that name are not this species. Although this tree at some stage was misidentified, in using the earlier name people are not carrying out identifications but simply continuing to use names as they have always done. This is different from misidentification, and it is a difference that is worth noting in considering the reasons why names change.

In summary, 'misapplication' refers to name usage. Misapplied names are, for the most part, incorrect names that have been used by a large number of people and are often names that have been perpetuated in publications. Misapplications begin with a misidentification no matter how excusable that might be. In contrast 'misidentification' refers to the act of giving an incorrect name to a plant.

Put simply, misapplication is the perpetuation of names resulting from an original misidentification; it is not the name itself that has changed, what has changed is the way it is used (its application). A misapplication requires a correct identification/determination.

Table 1 summarises the ways plant names are changed.

What name to use?

Horticulturists do not have to become botanical experts in order to know what name to use. There are published lists, either in hard copy or on the

Table 1: Name changes – summary table

Type of change	Examples
Nomenclatural changes	The earlier published name *Phragmites australis* has been adopted for what was once widely known as *Phragmites communis*. *Phragmites communis* is therefore now a synonym of *P. australis*.
	Rhododendron 'Sherbrook', registered with the International Cultivar Registration Authority (ICRA) for Rhododendron in 1983, was corrected to *R*. 'Sherbrooke' when it was realised that the name of the place after which the cultivar was named has the latter spelling.
	An intergeneric hybrid name must consist of a combination of part of the names of its constituent genera. The transfer of *Chamaecyparis nootkatensis* to the genus *Xanthocyparis* as *Xanthocyparis nootkatensis* has resulted in the new botanical name for the Leyland Cypress, X*Cuprocyparis leylandii*.
Taxonomic changes	The transfer of *Chamaecyparis nootkatensis* to the genus *Xanthocyparis* has resulted in the new name *Xanthocyparis nootkatensis*.
	The split of the species *Eucalyptus lehmannii* into two species *E. lehmannii* and *E. conferruminata*.
Misidentification	Mrs Brown's neighbour identified her Alister Clark-bred rose as *Rosa* 'Red Beauty' until she visited Alister's scrupulously labelled garden and saw that he had labelled it *Rosa* 'Red Queen'. Checking her rose again against pictures and a description she realised that she had made a misidentification.
Misapplied name in horticulture	The name *Pyrus ussuriensis* has long been used in the nursery industry for many plants that are now known to be *Pyrus calleryana*.
	Brunsvigia × *tubergenii* is a listed name of no botanical standing for X*Amarygia parkeri*.
Misapplied name in botany	A number of years ago it was found that the type specimens of alpine bottlebrush, *Callistemon sieberi*, housed in a herbarium in Geneva did not resemble plants known as *C. sieberi* in Australia but resembled what was then known as the river bottlebrush, *C. paludosus*. This meant that the name *C. sieberi* as used in Australia for many years, had been misapplied.

internet, that can be checked. **Part 4 – Plant name resources**, has listings of books and websites to help with the checking of plant names.

If you are still unsure about a name after consulting these resources, contact your state or regional botanic garden for advice. Remember though, that checking the name of a plant on a list is not the same as checking the plant's identification. You might have an up-to-date name for the wrong plant!

Although name changes can be exasperating, appear nit-picking, or even seem totally unnecessary, the internationally accepted rules have provided a framework in which botanists can satisfactorily resolve nomenclatural problems. In addition, names resulting from taxonomic research have given us much greater insight into plant relationships and the evolutionary history of the Plant Kingdom.

cultivated

Part two

plants and cultigens

The *International Code of Nomenclature for Cultivated Plants*

Giving special names to human-altered and specially selected plant variants dates back to ancient times. It is likely that deliberate plant selection began about 10 000 years ago when plants and animals were first domesticated as humans gradually changed from the nomadic hunter–gatherer lifestyle to living in settled communities. This was the dawn of modern agriculture and horticulture and it seems likely that names would have been given to selections of food crops made from this period on.

Written records of specially selected garden plants date back to about 160 BC when economic crops like apples, figs and olives were given the names of the sites where the propagation material was collected (Stearn 1986).

In the 19th century such plants, and some hybrids, were given horticultural names, sometimes in Latin, and these were added to the existing botanical names in an informal way and loosely categorised as 'forms', 'garden varieties', 'garden hybrid varieties', 'races', 'strains', etc. These names may have assisted communication but they could, nevertheless, be lengthy or difficult as in *Canna* Grossherzog Ernst Ludwig Von Hessen, or *Phlox drummondii nana compacta punicea striata*. So, at the turn of the 20th century, names used in nurseries were either in Latin form, often without proper typification, or in English (or other vernacular).

In the early 20th century, an effort was made, especially by the Bailey Hortorium in the United States of America, to describe specially selected and bred plants according to the prescriptions of the *Botanical Code*, but this proved unsatisfactory.

These preliminary steps in horticultural nomenclature eventually led, in 1953, to the publication of the *International Code of Nomenclature for Cultivated Plants* (Stearn 1953) with the later 1958 edition extending the horticultural focus of the first edition to include the plants of agriculture and forestry (Fletcher *et al.* 1958). There have now been seven editions of the *Cultivated Plant Code*, the latest being that of 2004 (Brickell *et al.* 2004; Figure 19).

Cultivated plants

In our gardens we have wild plants that have botanical names applied according to the rules of the *Botanical Code* (e.g. *Quercus robur,* English Oak and *Callistemon salignus,* Willow Bottlebrush). We also have plants which have arisen by deliberate hybridisation, by accidental hybridisation in cultivation, by selection from existing cultivated stock or as selections from variants within wild populations maintained as recognisable entities solely by continued propagation (*Cultivated Plant Code*; Trehane *et al.* 1995).

The *Cultivated Plant Code* governs the names of this second assortment of plants, which has been defined as 'distinguishable groups of plants whose origin or selection is primarily due to the intentional actions of mankind' (*Cultivated Plant Code*; Brickell *et al.* 2004).

Figure 19:
International Code of Nomenclature for Cultivated Plants (2004)

The cultigen

Surprisingly, except for the confusing and ambiguous terms 'wild' and 'cultivated' we have no widely accepted collective terms for the above two groupings. The word 'cultigen' is a useful term for what can, for simplicity, be called human-altered plants, including all plants whose names are in part governed by the *Cultivated Plant Code*. It also helps clarify the confusing difference between the expression 'cultivated plants' of everyday speech (meaning plants in cultivation) and the cultivated plants of the *Cultivated Plant Code*. From now on, we will refer to the plants covered by the *Cultivated Plant Code* as cultigens, and to botanical names with a component governed by the *Cultivated Plant Code* as cultigen names.

Humans have altered, in various ways, plants that originally grew 'untouched' in the wild (see Table 2). As this now influences the way in

Table 2: Broad groups of cultigens

Cultigen	Description
Modern cultigens produced by breeding and selection	This is by far the largest group of cultigens ranging from selections of unusual minor variants that have arisen in cultivation and simple hybrids, to plants that are the result of highly sophisticated breeding and selection programs involving many species over a period of many years. These modern cultigens are mostly recognised as cultivars and given a cultivar epithet (see Table 3 for the different kinds of cultivars).
Ancient cultigens	A small group of plants (often precursors of important economic crops) that occur in the wild but which have undergone selection and distribution by humans for so long that their original ancestral distributions and forms in the wild are uncertain or unknown. It is an historical anomaly that many of these plants were given names (binomials) under the *Botanical Code* before the *Cultivated Plant Code* was introduced. They have retained these names even though, nowadays, they would be given additional names under the *Cultivated Plant Code*; for example, *Zea mays*, Corn; *Solanum tuberosum*, Potato.
Graft-chimaeras	A small group of plants grafted to produce mixed tissue (graft-chimaeras). The graft material may be from wild plants, special selections, or hybrids.
Genetically modified organisms	Plants produced by genetic engineering which are the result of the deliberate implantation of genetic material from different organisms.
Naturalised cultigens	This group is included here to emphasise that a cultigen is not defined by where it is growing; it remains a cultigen even when it is a garden escape.

which these plants are named, it will be beneficial to look more closely at the major kinds of cultigen.

HISTORY OF CULTIGEN NOMENCLATURE

The use of Latin for human-derived plants has always been a source of frustration. The German botanist Karl Koch as early as 1865 considered Latin names for garden forms a source of confusion (Stearn 1986). The American, Liberty Hyde Bailey, was an early pioneer of horticultural taxonomy and in preparing his *Manual of Cultivated Plants* (an identification guide to the plants cultivated in North America that was published in 1924), he suggested that changes in botanical nomenclature would be desirable for people with 'a scientific interest in cultivated plants'. Bailey recognised the strengths of binomial nomenclature, Latin names and the Linnaean classification hierarchy and did not want to upset it. However, he was having problems putting domesticated plants, often the products of long, complex and unknown breeding and selection programs, into the Linnaean pigeonholes. Many simply did not fit neatly into the categories of species, variety and so on. In 1918, he decided to retain the Linnaean system of binomials but to refer to species-like domesticated plants as 'cultigens' and they would have a type, name and description in accordance with the *Botanical Code*. This word coined by Bailey was contrasted with his word 'indigen', the latter being Bailey's word for wild plants which, of course, could be brought into cultivation as cultivated indigens. Bailey's original meaning of the word cultigen has, over the years, changed from 'species-like plant that has arisen under domestication' to the present-day understanding as used in this book. However, it was clear to Bailey that many cultigens were more like botanical varieties than species and so, five years later in 1923, he established a classification category for these plants. He used the word 'cultivar' as an abbreviation of the words 'cultivated variety'. To fit in with the Linnaean system, he described these in the same way as botanical varieties according to the *Botanical Code* and his use of this word included what we would now call Groups. The designation 'group', has been used since at least the first *Cultivated Plant Code* was compiled and was formally introduced in the 1958 *Cultivated Plant Code* to designate assemblages of similar cultivars.

Which plants and which names are covered by which Code?

In general terms, the *Botanical Code* deals with the Latin part of *all* plant names. Examples might be *Solanum tuberosum* and *Camellia japonica*. The *Cultivated Plant Code* then deals with the non-Latin additional names given to those cultigens considered sufficiently important to warrant their own name. In the names *Solanum tuberosum* 'King Edward', *Solanum tuberosum* Red-skinned Group and *Camellia japonica* 'Emperor Wilhelm', it is the *Cultivated Plant Code* that governs the use of the names 'King Edward', Red-skinned Group and 'Emperor Wilhelm'.

There are a few minor exceptions to this general rule and these are included in Table 9 of the Appendix, which summarises in detail how the *Botanical Code* and *Cultivated Plant Code* are applied.

Cultigens and the Cultivated Plant Code

It would be extremely convenient if the *Botanical Code* dealt with all wild plants and the *Cultivated Plant Code* covered all cultigens. Unfortunately, the situation is not so clear-cut.

Firstly, the *Cultivated Plant Code* is a practical document. It is concerned with particular plants that are considered to be sufficiently distinct and useful for human purposes to merit a special name in addition to the usual Latin botanical name; it is not concerned with naming every single variation that might arise in cultivation. William Stearn, who compiled the first *Cultivated Plant Code* in 1953, clearly had this in mind when he defined the 'cultivated plants' of the *Cultivated Plant Code* as:

> ... plants raised in cultivation which differ significantly from their wild ancestors or, if taken into cultivation from the wild, are worthy enough of distinction from wild populations for horticultural purposes to merit special names (Stearn 1986).

Secondly, a cultigen name consists of a Latin component governed by the *Botanical Code* together with the cultivar and/or Group epithets governed by the *Cultivated Plant Code*. So a part of all cultigen names is governed by the *Botanical Code* (there are a few minor exceptions to this rule, see Table 9).

Thirdly, there are some ancient and modern cultigens, and some hybrids, given names under the *Botanical Code* rather than the *Cultivated Plant Code* (see Table 9).

In spite of these exceptions it is clear that almost all cultigens have part of their full scientific names that fall under the *Cultivated Plant Code*.

THOSE CULTIGENS GIVEN NAMES UNDER THE *CULTIVATED PLANT CODE* FULFIL THREE CRITERIA

1   They have special features of sufficient importance to warrant a name under the *Cultivated Plant Code*.
2   The special features are the result of deliberate human breeding and/or selection, and do not occur in wild populations (except in the few cases where they represent a desirable part of the natural variation found in wild populations that is not covered by a botanical name).
3   It is possible to reliably perpetuate the special features in cultivation.

Principles of the Cultivated Plant Code

The *Cultivated Plant Code* follows the same format as the *Botanical Code*: a set of Principles, Rules and Recommendations. It also has a similar function to the *Botanical Code* as it aims to provide uniformity, accuracy and stability in the naming of cultigens, but it is much shorter than the *Botanical Code*. It is also administered by a different body, the *International Union of Biological Sciences Commission for the Nomenclature of Cultivated Plants*.

The Principles at the beginning point out that the nomenclature of cultivars and Groups is governed by the *Cultivated Plant Code* and:

*   is based on priority of publication (except in specified cases)
*   that these names are freely available for use and not to be confused with marketing names and common names, neither of which are controlled by the *Cultivated Plant Code*
*   that to stabilise the application of names the registration of cultivar and Group names by International Cultivar Registration Authorities is to be encouraged, as is the preparation of Nomenclatural Standards (herbarium specimen or other items which define a cultivar or Group, see page 65) and
*   that the *Cultivated Plant Code* is produced and promulgated by those concerned with the accurate naming of cultigens, but it has no legal force.

Cultigen classification

Cultigen classification is slightly different from that of wild plants.

THE CULTIGEN HIERARCHY

As we have seen in Table 2 (page 47), cultigens are a diverse assemblage of plants including simple hybrids between two species in cultivation; complex hybrids that are the result of long breeding programs involving many species; genetically modified plants (i.e. with genetic material artificially implanted); graft-chimaeras, and so on. Arranging cultivars into a multi-level nested hierarchy putting botanically closely related plants into the same groups in the same way as the *Botanical Code* would be extremely difficult.

RANKS

In the Linnaean nested (boxes-within-boxes) hierarchy of the *Botanical Code*, a plant can only belong to one taxonomic grouping at any rank so, for example, one particular plant species cannot be a member of two genera in the same classification scheme.

For cultigens covered by the *Cultivated Plant Code*, there are just two possible ranks, the **cultivar** and the **Group**, and it is possible for a particular cultivar to be a member of several Groups. For example, in cultigen classification, a large group of purple dissected-leaved maples may be given a Group name and the rest left undefined. Some of these purple dissected-leaved maples might belong to another Group based on the habit of the plant and some, possibly the same ones, might belong to a smaller Group with extremely finely dissected leaves.

In cultigen classification, we might also wish to group plants that are botanically unrelated, say those in different species with crinkled leaves or long flower stalks. Thus it is possible for a particular cultivar to belong to several different Groups based on different criteria.

So, in practice, cultivars are grouped in any way that is useful, or if there is no benefit, a Group will not be defined. Even so, cultivars like *Agonis flexuosa* 'Variegata' will still fit into a botanical hierarchy.

It is clear from the above that the classification of cultigens is different from that of wild plants. Nevertheless, since the Group is a more inclusive rank than the cultivar, the classification system can still be regarded as hierarchical.

The *Cultivated Plant Code* defines the **cultivar** (as a rank) as:

… the primary category of cultivated plants whose nomenclature is governed by the *Cultivated Plant Code*.

The *Cultivated Plant Code* defines the **Group** (as a rank) as:

… a formal category for assembling cultivars, individual plants or assemblages of plants on the basis of defined similarity.

The reasons for establishing and maintaining a Group will vary according to the needs of particular plant users.

TAXA

Taxa of cultigens differ from the taxa of wild plants in several ways, apart from being given cultivar and/or Group names:

- They are the result of artificial selection and are maintained essentially in artificial habitats, not natural ones (although they may escape into the wild).
- The means of fixing their names by using Nomenclatural Standards, though serving the same purpose, is different from the type system used for wild plants.
- Taxonomic botanists tend to avoid cultigens: they separate them from wild plants in herbaria and exempt them from revisions and evolutionary studies.
- The placing of cultivars into Groups is not necessarily done using botanical criteria.

For these and other reasons it may be argued that the word 'taxon' is not appropriate for cultigen groupings, and some workers prefer to use the word 'culton' (an abbreviation of 'cultivated taxon'; pl. *culta*) to indicate a distinguishable group of cultigens (Hetterscheid 1994).

The latest 2004 *Cultivated Plant Code* (Brickell *et al.* 2004) uses the word 'category' instead of 'rank', and 'distinguishable groups of cultivated plants' instead of 'taxon'. This draws attention to the differences discussed above and also avoids technical words. However, the new expressions present their own difficulties, not the least of which is the common usage of the word category in a different sense from that intended in the *Cultivated Plant Code*, and the wordiness of the expression 'distinguishable group of cultivated plants'. In the 2004 *Cultivated Plant Code* only two ranks (called categories) are used for cultigens: the cultivar and the Group.

Since the Group is more inclusive than the cultivar, in a similar fashion to the genus being more inclusive than the species, we consider the word 'rank' to be satisfactory here. We also recognise that there are significant differences between the concepts of a 'taxon' and a 'culton' but retain the former for its precision and similar usage in the classification of both wild

Figure 20: *Canna × generalis* is a popular hybrid with many cultivars

Image: Royal Botanic Gardens Melbourne

plants and cultigens, and in line with the definition of a taxon in the *Botanical Code* as 'a group into which a number of similar individuals may be classified'. The words 'rank' and 'taxa' serve the same purpose in the classification of both wild plants and cultigens, and we consider are best retained for both situations with the optional use of the word 'culton' for a 'cultivated taxon' (or 'cultigen taxon').

Kinds of cultigen

CULTIVARS

Many popular garden plants such as roses, camellias, cannas and grevilleas have been bred or selected for ornamental features: generally unusual or bright foliage and flower colours, interesting flower shapes, and appealing forms (Figure 20). There are also many economically important plants in agriculture and forestry that have been deliberately selected or bred for features such as increased yield, better flavour, greater resistance to disease, and so on. These are referred to as cultivated varieties or 'cultivars' for short.

The essential feature of cultivars is that they have distinct and desirable characteristics that can be reproduced reliably and maintained in cultivation.

The *Cultivated Plant Code* defines the cultivar (as a taxon, not a rank) as:

> … an assemblage of plants that has been selected for a particular attribute or combination of attributes and that is clearly distinct, uniform, and stable in these characteristics and that, when propagated by appropriate means, retains those characteristics.

How are cultivars different from cultigens? Cultigen is a more broadly encompassing idea than 'cultivar' and denotes *all* human-altered plants, while 'cultivar' is a technical taxonomic term denoting either a cultigen rank, or cultigen taxon.

THE DIFFERENT KINDS OF CULTIVAR

Cultivars are of many different kinds and they may be propagated in several ways: they are not necessarily genetically identical, and occasionally the character that has been selected in the cultivar may not be visible, as is the case with rust-resistant poplars (Table 3). Here are examples of the more common sorts of cultivars:

- Common garden plants propagated vegetatively by division, cuttings, grafting or budding; for example, *Rhododendron* 'Alarm' or *Malus* 'Gorgeous'. In this group would be most of the garden colour and habit variants of rhododendrons, camellias, roses, and so on. These may have arisen as sports or mutations; for example, variegated leaves or a colour sport of a flower on a camellia. Alternatively, they might be a particular selection from the offspring of a hybrid cross. A genetically uniform group of individuals propagated in this way may be referred to as a clone.
- Seed produced by controlled crossing. These cultivars are not necessarily genetically uniform, but they do have shared characteristics that define the cultivar. Most annual bedding plants and garden vegetables are of this sort; for example, Tomato 'Apollo'. Annual bedding plants that are obviously not identical but which may share some character, such as the form of the petals or stature, are also given cultivar names. Examples are *Verbena* 'Flagship', *Portulaca* 'Sunnybank', or a mixed colour selection of petunias.
- Some vegetable cultivars are vigorous or high-yielding F1 (first generation) hybrids that result from crossing stable inbred parental lines, and that do not breed true in subsequent generations. However, provided they can be formed from the original parental stocks anew each time, a cultivar name is appropriate; for example, Broccoli 'Skiff'.
- Cultivar names are also given to particular selections of wild-growing plants (see Provenance, page 59); for example, *Helichrysum* 'Diamond Head'.

Table 3: Different kinds of cultivars as detailed in the *Cultivated Plant Code*

CULTIVARS	
All cultivars have features that are distinct, uniform, and stable under propagation	
CLONES Produced by asexual propagation from any part of a plant, e.g. divisions, cuttings, grafts, budding	
Topophysic	from a particular plant part, e.g. lateral branch
Cyclophysic	from a particular phase of the growth cycle, e.g. juvenile leaf
Aberrant growth	e.g. witches brooms
Graft-chimaeras	
SEED-PRODUCED	
Uncontrolled pollination	when it can be distinguished by one or more characters that are distinct, uniform and stable under propagation
Lines	produced by repeated self-fertilisation or inbreeding
Multilines	made up of several closely related lines
F1 hybrids	the result of a deliberate repeatable single cross between two pure lines
Sourced seed	plants grown from seed of a particular provenance (topovariant)
GENETICALLY MODIFIED PLANTS Plants containing implanted genetic material from different germplasm	

GROUPS

The *Cultivated Plant Code* defines the Group as:

> ... a formal category for assembling cultivars, individual plants or assemblages of plants on the basis of a defined similarity ... according to the required purposes of particular users.

Examples are: *Rhododendron boothii* Mishimiense Group, *Hosta* Fortunei Group, *Prunus* Sato-zakura Group (Figure 21), and *Solanum tuberosum* (Red-skinned Group) 'Desiree'.

Although not well known to most horticulturists, the category Group is a convenient way of naming cultivars with particular characteristics in common.

GRAFT-CHIMAERAS

The graft-chimaera is fairly rare in horticulture but is a plant consisting of tissue from two or more taxa placed in close association by grafting.

Figure 21: The cultivar *Prunus* 'Fugenzo' belongs to the *Prunus* Sato-zakura Group, the Japanese Flowering Cherries.

Image from: Miyoshi M (1916). Die Japanischen Bergkirschen, ihre wildformen und kulturrassen. *Journal of the College of Science, Tokyo Imperial University* 34, 1–175.

CULTIGEN HYBRIDS

Hybrids occur both in the wild (see Natural hybrids, page 27) and in cultivation. Indeed, the same crosses may even occur in both situations. Hybrids in cultivation are generally produced by deliberate crossing, usually between species, to produce interspecific hybrids. For example, the hybrid *Abelia*, which is a cross between *Abelia chinensis* and *Abelia uniflora*, is named either *Abelia chinensis* × *A. uniflora* using what is referred to as a hybrid formula. Alternatively, it can be called *Abelia* ×*grandiflora*, which is a collective name covering all offspring when hybridising these two species. The name is formed according to the procedure stipulated by the *Botanical Code* for naming a new species. The multiplication sign indicates that the plant is of hybrid origin.

It is important to note that a name like *Abelia* ×*grandiflora* applies to all the offspring of this particular cross whenever it occurs, and it can be made numerous times over a long period. The offspring of crosses like this one may be extremely variable, showing a mix of the characteristics of the two parents, so those offspring that have attributes of value to horticulture are often perpetuated by vegetative propagation and given cultivar names such as *Abelia* × *grandiflora* 'Francis Mason' (Figure 22).

As we have seen in **Part 1**, describing a new species entails a fairly elaborate procedure involving the use of Latin, publication in scientific journals, preparation of a type specimen and so on. Horticulturists are

Figure 22: *Abelia × grandiflora* 'Francis Mason' is yellow-leaved selection of the hybrid cross between *A. chinensis* and *A. uniflora*.
Image: Royal Botanic Gardens Melbourne

generally not prepared to do this, nor do they like writing out a lengthy hybrid formula which seems an unnecessary complication. In many cases, a selection is simply made of one of the more appealing offspring of the cross and this is given a cultivar name that is placed directly after the genus name, so there is no hybrid species epithet. This is completely acceptable under the *Cultivated Plant Code*. In fact, the absence of a specific epithet in a cultigen name is a good indication that the plant is of hybrid origin. The cultivars *Camellia* 'Donation' and *Grevillea* 'Rachel' are selections from the offspring of hybrid crosses.

Hybrids between species of different genera will give rise to hybrid genera (intergeneric hybrids) that may also be named as, for example, ×*Chamaecyparis leylandii*, Leyland's Cypress, a cross between *Cupressus macrocarpa* and *Chamaecyparis nootkatensis*. Being an intergeneric hybrid, the multiplication sign is put in front of the genus name, which is generally a combination of the names or parts of the names of the genera from which it is derived (known as a condensed formula). In recent times, *Chamaecyparis nootkatensis* was placed in the genus *Xanthocyparis*. This meant that a new intergeneric hybrid name was required for the Leyland Cypress and ×*Cuprocyparis leylandii* was chosen. Special selections are given cultivar names as in ×*Cuprocyparis leylandii* 'Leighton Green'.

Hybrids involving multiple genera do occur, especially in horticulture, and the orchids provide good examples. Hybrids involving three genera may use a condensed formula formed either as above (e.g. ×*Sophrolaeliocattleya* = *Sophronitis* × *Laelia* × *Cattleya*), or from a person's name ending in *-ara* (e.g. ×*Devereauxara* = *Ascocentrum* × *Phalaenopsis* × *Vanda*). For hybrids of four or more genera, only a person's name ending in *-ara* may be used. None of these condensed formula names may consist of more than eight syllables.

The name of a hybrid genus must be validly published (established) in a journal such as *Orchid Review* with a statement of the parent genera but no description, diagnosis or nomenclatural type is necessary. However, names of hybrid genera appearing in trade catalogues or non-scientific newspapers after 1 January 1953, or in seed exchange lists after 1 January 1973, are not validly published (established) in these publications.

Naming wild plants brought into cultivation

How do we name plants that are brought out of the wild directly into gardens? As these plants are not cultigens (they have not been altered in any way by humans) then it might be expected that they would simply retain their botanical names, and in almost all cases this is so, but there are a few minor exceptions.

The case of wild plants in cultivation demonstrates how the naming of plants in cultivation is, on occasion, a matter of expediency: it provides the means for giving a name when one is wanted, regardless of the origin of the plant.

WILD PLANTS IN CULTIVATION NAMED UNDER THE *BOTANICAL CODE* ONLY
Some garden plants have been selected directly from wild populations without undergoing any genetic alteration by humans. Examples would be common park and garden trees like *Betula pendula*, *Eucalyptus nicholii* and *Liquidambar styraciflua*. These plants were initially propagated by seed, cuttings or scions obtained from plants growing in the wild, and their phenotypes (observable characteristics) are generally representative of those found in wild populations even though they represent only a small part of the natural genetic variation of the species. Usually, after the initial wild collection, these plants are re-propagated from the trees in cultivation rather than by re-collecting material from the wild. Nevertheless, they are considered, to all intents and purposes, the same as their wild counterparts as they have no special characteristics requiring a name under the *Cultivated Plant Code*, even though they were deliberately collected from nature and

are now growing in artificial environments. They are therefore given names under the *Botanical Code* only.

In forestry, when tree seed is collected from a natural population to establish new forests or plantations, its geographical origin is recorded. Seed from a particular area may be suited, for example, to grow in saline or flooded environments. The term 'provenance' is used to describe the geographical origin of the population from which the seed was collected. This is written, for example, as follows: *Eucalyptus regnans* (Mt Donna Buang). This is a convenient naming system for forestry variants when a cultivar name is not considered appropriate.

WILD PLANTS IN CULTIVATION THAT ARE GIVEN CULTIVAR NAMES

Sometimes horticulturists notice unusual variants in the wild that have special ornamental appeal. For example, a plant that usually has purple flowers might occasionally produce attractive double white ones, and it may be possible to reproduce the double white flowers reliably in cultivation. The botanical description of this plant might state that the flowers may be either white or purple, so the double white flowers are simply part of the acknowledged natural genetic variation of the species. In many cases like this botanists do not regard differences in flower colour of sufficient importance to warrant recognition by being given a special Latin name. Although colour may be insignificant botanically, it can be extremely important horticulturally.

In cases like this where a name is needed in horticulture to mark the selection of a botanically unnamed part of natural variation, then a cultivar name is given.

Some further examples will help illustrate how this situation might occur.

In a revision of *Rhododendron* over 20 years ago a number of 'species' familiar to gardeners were combined because botanists did not regard the differences between them as being of botanical significance. Unfortunately for horticulturists, many of these species were of considerable interest to horticulture but, as a result of this work, no longer had botanical names. To ensure that names remained to recognise these plants in horticulture they were given Group names so, for example, *Rhododendron desquamatum* became, eventually, *R. rubiginosum* Desquamatum Group.

The plant *Mahonia japonica* is regarded by some as a cultigen and therefore best recognised as such by being given a cultivar epithet. In a case like this, the *Cultivated Plant Code* allows the name to be written as *Mahonia* 'Japonica'.

The tree *Liquidambar styraciflua*, as part of its natural variation, displays spectacular autumn foliage colours ranging from oranges to red and deep maroon; sometimes all colours are on the same tree as a complex variegation, but at other times one colour dominates. It is possible, by budding or apical grafting, to clone particularly appealing colours or colour combinations. It is quite likely that these particular colours occur in nature, so the clones may represent part of the plant's natural variation even though they are specially selected and propagated. However, only a small, particularly appealing part of this natural variation has been selected for horticulture and this has no distinguishing name under the *Botanical Code*, so a name is given under the *Cultivated Plant Code*, say *Liquidambar styraciflua* 'Tirriki', which has deep red autumnal colour.

Cornus florida f. *rubra* occurs occasionally in wild populations; it has red bracts instead of the usual white ones. When brought into cultivation these plants are sometimes called *Cornus florida* 'Rubra'. The American book *Hortus Third* uses the cultivar notation for many natural varieties and forms. However, the bract colours are not always identical across all plants and several clones may be present in cultivation, each originating from a particular wild collection. Since a feature of a cultivar is that it can be propagated without change, there is a reasonable expectation that bract colour would remain constant in plants named *Cornus florida* 'Rubra'. Thus, in this case there is good cause for retaining the name *C. florida* f. *rubra*, which allows for the variation found in wild populations. If we wanted to recognise one particular bright red clone, we could name it, say, *Cornus florida* f. *rubra* 'Scarlet Wonder'. With a name like this, nursery labels for simplicity would probably omit the f. *rubra*. Having said this, it should nevertheless be noted that cultivars need not necessarily be clones, nor need they be uniform, and some variation may be permitted within their descriptions.

Many cultivars of Australian native plants are simply selections from the wild; for example, *Callistemon pallidus* 'Mount Oberon', which is a large-flowered variant from Mount Oberon at Wilsons Promontory in southern Victoria. If these plants continue to show the features that were the reason for the selection then the cultivar epithet is entirely suitable. However, vegetative propagation is usually required and if the plants on sale exhibit considerable variation, usually as the result of propagation from seed, then they are not by definition the same cultivar.

There may be disputes about how much, if at all, a plant has been altered before it may be given legal protection under Plant Breeder's Rights (PBR) legislation. In Australia at least, the PBR legislation defers to the

Cultivated Plant Code in its interpretation of which wild selections may be given a cultivar name.

Wild plants named separately by botanists and horticulturists
Within wild populations of *Eucalyptus caesia* are robust plants with attractive pendulous branches, large leaves and an exceptionally white waxy bloom on the twigs, fruits and flower buds. Selections of this highly ornamental variation were made for horticulture and given the name *E. caesia* 'Silver Princess'. Later, this natural variation was acknowledged by botanists with the name *E. caesia* subsp. *magna* under the *Botanical Code*. Although the cultivar epithet 'Silver Princess' persists in the nursery industry, it no longer represents botanically unnamed genetic variation. In a case like this the name *E. caesia* subsp. *magna* is preferred. It is worth noting that if subspecies *magna* had been named first, and then subsequently reclassified as a cultivar, then it would have been acceptable, under the *Cultivated Plant Code*, to convert the subspecies epithet to a cultivar epithet as *E. caesia* 'Magna'.

The name *Fagus sylvatica* Cuprea Group may be used to include the range of plants in cultivation that have copper-coloured leaves. However, in the past, wild plants with coppery-purple leaves were placed under the name *Fagus sylvatica* f. *purpurea*. It therefore appears possible, on rare occasions, for a plant to have independent and equally acceptable botanical and cultigen names, one under each *Code*, serving the needs of scientists on the one hand and horticulturists on the other. The *Cultivated Plant Code* recommends that where plants meet the criteria of being recognised as cultivars or Groups, then they should be named in accordance with the provisions of the *Cultivated Plant Code*, not under the provisions of the *Botanical Code*.

Publishing cultigen names

Like the publication of new species of wild plants, the publication of new cultigen names has precise requirements. To be formally recognised, the new taxon must be *published*, *established* and *accepted*.

Publication
Under the *Cultivated Plant Code* publication is achieved essentially by distribution of printed material in the public domain and dated to at least the year, and preferably sent to the appropriate International Cultivar Registration Authority (ICRA), if one exists for the particular genus (see

Figure 23: Dated nursery catalogues are suitable for publishing new cultivar names
Image: Royal Botanic Gardens Melbourne

Part 4 – Plant name resources). A dated trade catalogue is usually sufficient (Figure 23).

ESTABLISHMENT

To be *established*, the name must, in general terms, also conform to the *Cultivated Plant Code's* nomenclatural rules and, if published after 1 January 1959, it must be accompanied by a description (or a reference to a previously published description) of the character(s) that distinguish it from other taxa. Ideally, an illustration would be included and other background information that might be of interest, including a herbarium specimen (see Nomenclatural Standards, page 65).

After this date any language, other than Latin, is valid. The *Cultivated Plant Code* recommends that a new cultivar name be accompanied by a full description, and the name(s) of the people who originated and described it. The plant description should include details of parentage (if appropriate), also the particular characteristic(s) that distinguish it from related cultivars.

ACCEPTANCE

An accepted name is the earliest established one for that taxon, unless an ICRA designates a later name that may have become more widely accepted. If no ICRA exists, a proposal must be submitted to the International Union of Biological Sciences Code Commission for a ruling.

FORMATION OF CULTIVAR AND GROUP EPITHETS

A cultigen name should consist of at least a genus name (or unambiguous common name) followed by a Group and/or cultivar epithet which, nowadays, is required to be in a language other than Latin.

Occasionally, the same cultivar name has been used for several species of the same genus. In cases like this, it is important to distinguish the cultivar further. For example *Juniperus* 'Variegata' is a name that may apply to several species of juniper, so the name must be written in full as *Juniperus chinensis* 'Variegata'.

Since 1 January 1996 the epithet must consist of no more than 30 characters and it is preferable that the epithet is fairly brief, and uncomplicated in construction or pronunciation.

Confusing cultivar names like *Rosa* 'Rose' and Plum 'Apricot', where the cultivar name is another name for the genus, may not be established. However, where names are of different genera then they are permitted, as in *Camellia* 'Rose', *Lilium* 'Erica' and *Magnolia* 'Daphne'. For the same reason, names including words like 'Variety', 'Var.' or 'Form' must be avoided, and since 1 January 1996, words like 'cultivar', 'grex', 'series', 'strain' and 'hybrid' must not be included. Other epithets to be avoided are names that might cause confusion through similar spelling to that of an existing name (e.g. 'Suzanne' and 'Susanne'), names that are misleading in any way (e.g. Apple 'Redskin' when the skin is not red), common words or expressions like 'Five Dollars', or epithets that might give offence.

Cultivar epithets may include apostrophes, commas, single exclamation marks, full-stops (periods), hyphens, forward and backward slashes, and numbers, but they must not contain question marks. They may also be a newly invented word or word combination, or even be in the form of a code of up to 10 characters, excluding spaces. Since 1 January 2004, a cultivar name must not consist of a single letter, or be composed only of Arabic or Roman numerals.

A Group epithet is a word or phrase with the word 'Group' added; for example, *Brassica rapa* Pak-Choi Group. It is recommended that Group epithets, like cultivar epithets, should not be misleading or confusing in any way.

USE OF LATIN FOR CULTIGEN EPITHETS

The *Cultivated Plant Code* specifies that a cultivar name published on or after 1 January 1959 should not be a botanical name in Latin form. Prior to 1959 there were, however, many variants that were given botanical names at the rank of either variety or form, and these are permitted to remain, even though many of them have been reclassified as cultivars. An example of a name of this sort pre-dating 1959 is *Thuja occidentalis* var. *aurea*, which is now generally written as *Thuja occidentalis* 'Aurea'. Since 1 January 1959 all new cultivar names must be written in a common language (not Latin); for example, *Callistemon pallidus* 'Father Christmas'. The rules and recommendations applying to Groups are, for the most part, the same as those applying to cultivars.

Plants named under the *Botanical Code* only and subsequently reclassified as cultivars where possible have their lowest ranking epithet converted to a cultivar epithet; for example, *Mahonia japonica* becomes *Mahonia* 'Japonica' and *Cedrus atlantica* f. *glauca* when reclassified as a cultivar becomes *Cedrus atlantica* 'Glauca'. Other recent changes include the recognition of hybrid cultigens *Opuntia ficus-indica*, which becomes *Opuntia* 'Ficus-indica', and *Bauhinia blakeana*, Hong Kong Orchid Tree, now *Bauhinia* 'Blakeana' as it is a hybrid cultigen between *B. purpurea* and *B. variegata*.

TRANSLATION, TRANSLITERATION AND TRANSCRIPTION

Translation (changing the words of one language into those of another) of cultivar names from their original language is not permitted. So, for example, translating the French cultivar name *Hibiscus syriacus* 'L'Oiseau Bleu' to *H. syriacus* 'Blue Bird' is not permitted. Naturally, nurseries will use translations on their labels because they are more user-friendly and encourage sales. If a translation is used then it should be written as a trade designation, not as a cultivar. Group names, provided they are not in Latin, may be translated.

Transliteration is the changing of words of one script to those of another. This is permitted when the Group or cultivar name can be converted character by character as in transliterating Russian to English.

The ideographic scripts of Chinese, Japanese and Korean cannot be converted letter by letter using transliteration. Instead, sounds ('spell sounds') are converted into the other script, in a process known as transcription. This is done using a standard recognised conversion system; for example, Mandarin Chinese idiographic characters are converted into a script using the Roman alphabet using the Pinyin system.

PRIORITY

Like the Principle of Priority for wild plants, the first published epithet is the one accepted. However, for cultigens, exceptions are made when a name that is contrary to the rules is widely known and has been accepted and published by the relevant ICRA and submitted to the International Society for Horticultural Science (ISHS) Commission for Nomenclature and Cultivar

Registration. Or, if it is not on a register, a separate submission has been made to the Commission. The Commission also has the authority to resolve disputes and conserve cultivar and Group names.

AUTHORS
A full citation of a botanical name requires the name of the author who established the taxon. For cultigen names this is not necessary, although it can be useful in tracing the history of the particular cultivar or Group. The name of the author attributed with the establishment of the cultigen name is placed after the epithet of the cultivar or Group. The author of a taxon registered by a statutory plant registration authority (such as the Australian PBR Office) is the person or organisation granted the rights.

Nomenclatural Standards

When a new cultivar is released it is generally accompanied by a very brief description (often only a few words) as part of the marketing publicity. This promotional material may be the only published information that exists and it frequently omits a clear and extended description, or information about how and when the plant arose, or what distinguishes this particular cultivar from other similar ones. Over time, any remaining knowledge may be lost. In contrast, those cultivars protected under PBR legislation are required under law to have a full description. In Australia, this description is published by the Australian PBR Office in the *Plant Varieties Journal* which, since 2004, has only been available online. It includes an historical profile of the cultivar with a diagnosis that pays close attention to the similarities and differences between this cultivar and similar closely related ones that are referred to as comparators. In the United Kingdom, the Plant Variety Rights Office publishes the *Plant Varieties & Seeds Gazette*, in Canada the PBR Office publishes the *Plant Varieties Journal*, in New Zealand the Plant Variety Rights Office publishes *The New Zealand Plant Variety Rights Journal*.

We have seen that in wild plants the name of a plant is fixed by the use of a type specimen lodged in a herbarium (see page 17). In view of the difficulties of identification and lack of information about cultivars, the 1994 *Cultivated Plant Code* introduced Standards (now called Nomenclatural Standards) that would serve a similar function. These Nomenclatural Standards are usually herbarium specimens that are stored in a specially marked folder called a portfolio, which puts together illustrations, colour chart references, the original publication details, an

NATIONAL HERBARIUM OF VICTORIA (MEL)
MELBOURNE AUSTRALIA

NAME: *Banksia spinulosa* 'Stumpy Gold'
FAMILY: *Proteaceae*
PLACE of COLLECTION: Merricks Nursery
Merricks Victoria

DATE of COLLECTION: 1988
COLLECTOR: Sally Debenham
ORIGINATOR: Richard Anderson, Merric
Nursery, Merricks, Victori

ORIGINAL CITATION: A.P.S 'Banksias for Syd
DATE of INTRODUCTION: 1993
PORTFOLIO PREPARATION: Sally Debenham
ILLUSTRATIONS: 2 photographs
2 scans
CHECKED BY: Roger Spencer and Michèle Adler
REPLICATES:
NOTES:

NATIONAL HERBARIUM OF
VICTORIA (MEL), AUSTRALIA

account of parentage and origin, and any other information of interest (Figure 24). Occasionally, the Standard is an illustration only. Herbaria that are known to be actively storing Standards are designated in an appendix of the *Cultivated Plant Code* (see also Table 4, page 72).

The denomination class and the replication of names

We have seen how for wild plants replication of names is avoided under the *Botanical Code* through Principles 3 and 4, which demand that within a particular classification system there can be only one name, and that if there are two identical names available (homonyms) the one to be accepted is based on priority of publication (see page 19). Obviously, having two genera or families with the same name in the Plant Kingdom would be confusing. For similar reasons, it is not permissable to have the same specific epithet within a particular genus. However, it is possible to have the same specific epithet in different genera so we have, for example, *Acacia rubida* and *Eucalyptus rubida*.

The *Cultivated Plant Code* deals with the problem of replication of names by not allowing the same names within a denomination class, which is a specially designated grouping of plants. The denomination class is almost always a genus (or hybrid genus), but there are a few exceptions. For example, the family Orchidaceae has eight denomination classes for hybrid groups within the family. The genus *Iris* has two denomination classes: one for bulbous plants and the other for non-bulbous ones.

New names for existing cultivars

Nurseries and promoters sometimes deliberately coin 'new' cultivar names for plants that already have perfectly good ones. An example is the name *Houttuynia cordata* COURT JESTER, a recent promotion for the 19th century cultivar *Houttuynia cordata* 'Variegata' (Figure 25). Such promotions are understandable but to be discouraged from the point of view of stable nomenclature. They can also involve a degree of misrepresentation.

A related and very difficult problem is what to call rediscovered old cultivars when the names have been lost. In some cases these can be

Figure 24: Nomenclatural standard of *Banksia spinulosa* 'Stumpy Gold' housed at the National Herbarium of Victoria
Image: Royal Botanic Gardens Melbourne

Figure 25: The 19th century cultivar *Houttuynia cordata* 'Variegata' has more recently been promoted as *Houttuynia cordata* COURT JESTER.
Image from: Robinson W (1889). *The English flower garden – style, position, and arrangement.* 2nd edition. John Murray, London.

accurately identified, as with many roses. In other cases an old cultivar name, whose description approximates the features of the plants in question, is used. This can lead to error but the alternative – giving a new name – can also cause problems. Sometimes the situation can be avoided by using a generalised description in brackets such as (dwarf pink), placed after the name.

Re-use of cultivar names is only allowed when, in the opinion of a cultivar registration authority, the cultivar in question is no longer in cultivation; not available as breeding material in any gene or seed bank; not known in the pedigree of other cultivars and has not been widely available at any time.

Procedure for introducing a new cultivar

We have described the formal requirements for publishing a new cultivar, but what exactly is the procedure for introducing a new find to commerce? The following account covers much of what has been dealt with already but in a more applied way.

Let's say you have discovered a new cultivar, either in the wild or in your garden or nursery. Should it be reported or registered in some way? How can it be officially recognised, and can it be propagated and supplied to the nursery industry for distribution? Is it possible to make money from the discovery and can the new find be legally protected?

We have prepared a flow diagram to indicate the options available to people in this situation and described the procedures that need to be gone

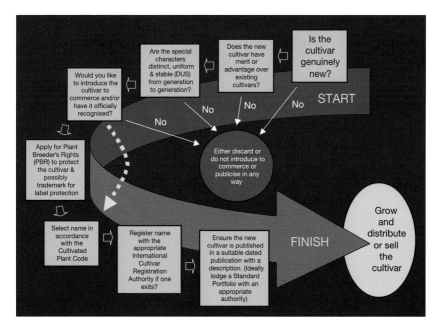

Figure 26: The steps required for introducing a new cultivar

through (Figure 26). More detailed information is noted in the points below.

IS THE PLANT GENUINELY NEW?

Ensure that you really do have something new. Check with knowledgable people in plant societies and the horticultural industry, then look through as much of the literature as possible. This will include specialist books on cultivars and the register of the appropriate ICRA, if there is one. A botanic garden should be able to help you with this part of the process. Remember that if cultivars are indistinguishable then they are regarded as the same, even if their origins are different.

DOES IT CLEARLY HAVE SOME MERIT OVER PLANTS ALREADY AVAILABLE?

Simply obtaining a 'new' plant is not in itself remarkable. Many plants in horticulture show seedling variation where plants grown from seed display a range of characters: English lavenders are notorious. Many new camellia cultivars have arisen as sports from existing cultivars, while the numerous foliage colour and habit variations to be seen in conifer nurseries are the result of chance sports in foliage. To be worthwhile, your new plant should show some clear advantage over its most similar relatives, so do not get caught up in a burst of initial enthusiasm until you are sure that you have something special, in ornamental appearance, crop yield, disease resistance or whatever. If there is any doubt about this, then you are entitled to let the market decide on the quality of your product, but it is best to be sure that you have something special before you release it.

Can the special characters that distinguish it be reproduced?

To be recognised as a new cultivar, you must be able to reproduce the special characters that make it unique. And to be acceptable, these characters must be distinct, uniform and stable ('DUS' for short) from generation to generation. To be absolutely sure of this it may be necessary to undergo quite prolonged trials. For example, dwarf conifers are known to sometimes suddenly revert to normal growth after several years and so trial periods of up to 25 years have been recommended for these.

Would you like to take economic advantage of the find?

You may not be able to take commercial advantage of your find directly as this may require a massive propagation program, good marketing and so on. Perhaps you could come to some arrangement with a grower. However, unless there is some legal protection, anyone who obtains your plant may reproduce it and sell it. More and more people are now taking advantage of PBR. If you wish to have legal protection for your plant then you can obtain PBR for a sum of money (necessary to cover the costs of trials to ensure that it is DUS, see above). Note that trade designations, like trademarks, might give you a marketing advantage but do not protect the cultivar itself. To take advantage of PBR you should contact the Plant Breeder's Right's Office in your country (see **Part 4 – Plant Breeder's Rights** for contact details) and they will supply you with all the details.

How do I choose a new name?

This should be quite easy but consult the section Formation of cultivar and Group epithets (page 63), which is briefly summarised here:
- Although several words can be used there is a maximum of 30 letters.
- As the name will be widely used it should be easy to spell and pronounce.

- It should not be ridiculously promotional; for example, 'Best Ever'.
- The words 'hybrid', 'variety' and 'improved' must be avoided, but 'cross' is permissable
- Keep punctuation simple, and generally avoid the names of other plants. *Rosa* 'Rose' is not permitted, although *Rosa* 'Iris' is allowed.
- The name should be unique for your group of plants, so you might have to check the appropriate international cultivar registers or ask a botanic garden to do this for you. Although by default the genus constitutes the denomination class, there are a few cases where a genus consists of more than one denomination class and this allows replication of names in that genus (see page 67).

ARE THERE ANY SPECIAL REQUIREMENTS FOR THE NEW CULTIVAR TO
BE OFFICIALLY RECOGNISED?

Even though you may not wish to take commercial advantage of your discovery, you should make sure that it is readily available as there is little point in putting on record a plant that nobody knows or grows. If your plant is from a group that has an ICRA (see Cultivar registration, below), then registering your plant will ensure that it is internationally recognised forever and this generally does not cost anything. Regardless of whether there is an ICRA, you should publish your name with a description and a picture of the new cultivar, preferably in a horticultural journal, although a publication like a distributed nursery catalogue would do provided it is dated to at least the year and there is a description of the characters that distinguish your plant from its most similar relatives. It is then said to be *published* and *established*.

In view of the enormous difficulties in recording cultivars, the new *Cultivated Plant Code* strongly recommends the creation of what is called a Standard Portfolio. This would be a herbarium specimen or illustration of the cultivar, a copy of the original description and any other information of interest. The Standard specimens can be housed in the herbaria listed in Table 4.

Cultivar registration

Large numbers of new cultivars are being constantly introduced to horticulture. How are these recorded and documented?

Before 1953 specialist societies for genera such as *Rhododendron*, *Iris* and *Camellia* controlled the naming and registration of new cultivars. For over 40 years now, however, a coordinated system has been operating, with

Table 4: Organisations maintaining Nomenclatural Standards

Country	Organisation	Location
Australia	Australian National Herbarium	Canberra, ACT
	National Herbarium of Victoria	Melbourne, Victoria
Canada	Royal Botanic Gardens	Hamilton, Ontario
The Netherlands	National Herbarium Nederland, Wageningen	Wageningen University Wageningen
New Zealand	Allan Herbarium, *Landcare Research* New Zealand	Lincoln
South Africa	National Herbarium, National Botanical Institute	Pretoria
United Kingdom	Royal Botanic Garden, Edinburgh	Edinburgh
	University of Reading (Plant Science Laboratories)	Reading, Berkshire
	Royal Horticultural Society	Wisley, Woking, Surrey
United States of America	Bishop Museum, Department of Natural Sciences	Honolulu, Hawaii
	Brooklyn Botanic Garden	Brooklyn, New York
	George Safford Torrey Herbarium, University of Connecticut	Storrs, Connecticut
	Caude E. Phillips Herbarium, Delaware State University	Dover, Delaware
	Willard Sherman Turrell Herbarium, Miami University	Oxford, Ohio
	United States National Arboretum	Washington, DC

ICRAs recording and publishing cultivar names (Figure 27). There are currently 71 ICRAs covering 3180 genera, but for many genera there is still no registration authority. It should be noted that registration is not an acknowledgement of the merit or distinctiveness of the cultivar or Group, but more a recording mechanism. ICRAs are mostly run by keen volunteers and should not be confused with statutory registration authorities like the PBR Office.

ICRAs are appointed by the International Society for Horticultural Science's Commission for Nomenclature and Cultivar Registration, and they follow closely the rules and recommendations in the *Cultivated Plant Code*. They are responsible for the registration of cultivar and Group names and they assist the stability of cultivar names by producing authoritative checklists and registers of all names known, past and present, for their

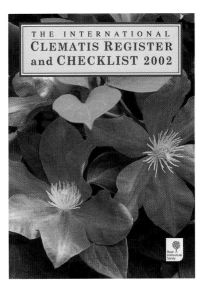

THE INTERNATIONAL
CLEMATIS REGISTER
and CHECKLIST 2002

Figure 27: New cultivar names are preferably registered with the appropriate International Cultivar Registration Authority

designated plants. They are also responsible for the publication and description of new cultivars within a particular genus. These authorities may be discriminating in their registration, but their basic function is to record new names. Names in genera with an international registration authority may be legitimately published independently of the authority, but this is not encouraged. Trademarks and trade designations may also be listed by an ICRA to provide reference links to true cultivar names, and thus minimise confusion.

Examples of organisations responsible for ICRAs are the International Camellia Society, the Royal New Zealand Institute of Horticulture, which is responsible for registering plant cultivars of selected New Zealand taxa, and the Australian Cultivar Registration Authority, which is responsible for registering cultivars of all Australian taxa.

The Australian Cultivar Registration Authority at the Australian National Botanic Gardens in Canberra maintains a collection of Nomenclatural Standards, slides and, where possible, living representatives. Descriptions of newly registered Australian native plant cultivars are published regularly in the journal *Australian Plants*. Keeping the necessary records may appear tedious but it is crucial for identification, and is a vital source of information for authors, breeders and horticultural historians.

There is a list of the addresses of registration authorities for particular genera in the *Cultivated Plant Code* and also on the internet (www.ishs.org/

icra/index.htm). **Part 4 – Plant Name Resources** also notes the ICRAs (see Tables 6, 7 and 8).

An application form obtained from the appropriate registration authority must be completed for registration. Failure to do this will make later name-checking and recording extremely difficult.

People describing or registering new cultivars should lodge a dried specimen in a public herbarium, with a coloured illustration or photograph. Places maintaining Nomenclatural Standards are noted in Appendix IV of the *Cultivated Plant Code* (see also Table 4).

Acceptance of a name for registration does not necessarily imply that the cultivar has merit. Producing new and different cultivars is often quite easy, so it is important that they should only be submitted for registration when they have ornamental or other characteristics that are unique and of special merit. This often requires careful selection and/or breeding over many years.

Marketing names (trade designations)

Over the last decade or so there has been a dramatic change in the kinds of names that are printed on retail nursery labels. Increasing competition and sophisticated marketing has resulted in a shift of emphasis from botanical and common names to legally protected marketing names. The modern nursery now uses similar methods to market plants as it does for other products, and that includes attractive packaging and catchy names. Plant labels have become promotional tools rather than simple name tags and are therefore designed to have instant sales appeal (Figure 28).

Acacia leprosa 'Scarlet Blaze'Ⓟ

FULL SUN   HEIGHT 5m   SPREAD 3m

FEATURES:
Striking blood-red flowers are produced from August to October. Ornamental, small evergreen tree. Fast growing with a semi weeping habit. More commonly known as Cinnamon Wattle as leaves crushed during warm weather release a cinnamon like fragrance.
SUGGESTED USES:
•Native gardens • Low maintenance areas
•Water wise gardening • Screening
WATER REQUIREMENTS:
Medium. Once established, quite dry tolerant requiring only occasional deep watering during extended periods of heat.
CARE:
Hardy and easy to grow. Tolerates frost and short periods of dryness, preferring a full sun to dappled shade position in well drained, moist soil.

A percentage of the revenue from sales of this plant will assist plant conservation programs of the Royal Botanic Gardens Melbourne

PBRⓅ Unauthorised commercial propagation or any sale, conditioning export, import or stocking of propagating material of this variety is an infringement under the Plant Breeder's Rights Act 1994.

Another introduction from PMA

www.pma.com.au

Figure 28: Labels are now an important component of plant marketing

75

The first words a customer sees on a nursery label are often a trademark. This may be a trademark of either the nursery itself (to reinforce the nursery brand and give it an edge over its competitors by encouraging people to return to that particular nursery), or of the grower that supplied the plant to the nursery (to increase sales of plants at the retail outlets supplied by that grower). Nowadays, when visiting garden centres in Australasia, Europe or North America, it is often impossible to tell what kinds of names we are dealing with when we look at the array of colourful display labels in front of us. Botanical names, if used at all, are frequently hidden in the small print on the back of the label. This situation seems likely to worsen as the numbers of legally protected promotional names mount up, and new kinds of legal protection for intellectual property arise.

Later we describe a way of distinguishing between the many different kinds of names that you will find on nursery labels.

Trade designations

Marketing names that are are not true scientific plant names are referred to by the *Cultivated Plant Code* as Trade Designations. There are two basic kinds of marketing names: PBR names and trademarks. Both of these kinds of names are used in several different ways so they require some explanation.

Plant Breeder's Rights

Growers have always been keen to protect their discoveries and inventions. If you have spent time and money breeding an exciting new cultivar then you would hope to reap the financial benefit that might arise as a result. But what is to stop someone obtaining, propagating and selling your new cultivar the moment you put it on the market?

In 1961, an international convention for the protection of new cultivars through intellectual property rights was adopted in Paris and this led to the formation of an intergovernmental organisation called the International Union for the Protection of New Varieties of Plants (UPOV) based in Geneva, Switzerland. The mission of UPOV is:

> To provide and promote an effective system of plant variety protection, with the aim of encouraging the development of new varieties of plants, for the benefit of society.

Where are Plant Breeder's Rights used?

The convention came into force on the 10 August 1968 with the European countries Britain, Germany, and the Netherlands forming the core membership (Denmark joined 2 months later in October). New Zealand and the United States became members in 1981, Australia in 1989, and Canada in 1991. Currently, there are 58 states that are party to the International Convention for the Protection of New Varieties of Plants. The first intergovernmental organisation to join was the European Community on 29 June 2005.

In the United States two systems are used to protect new plant varieties:

- plants that are sexually reproduced by seed or are tuber-propagated are protected through the Plant Variety Protection Office of the USDA under the Plant Variety Protection Act (amended in 1994), which complies with the 1991 Act of UPOV, and
- plants that are propagated asexually are protected through the US Patent and Trademark Office.

It has been possible to award plant patents since 1930 in the United States, but only for asexually propagated material. It was only when European countries enacted PBR legislation in the 1960s and were able to demonstrate that sexually reproduced plants can be uniform and stable that the US introduced the Plant Variety Protection Act in 1970. The Act is administered by the United States Department of Agriculture rather than the Patent and Trademark Office, and has never protected asexually propagated plants, the protection of which has remained with the Patent and Trademark Office. The plant patents awarded through the Patent and Trademark Office are similar to PBR, but the system is not aligned with the UPOV International Convention for the Protection of New Varieties of Plants. Like PBR in other countries, plant patents can only be awarded to a new variety that has not been sold or released in the United States for more than a year. Plant patents expire after 20 years from the date of filing the patent application, which is similar to PBR in Australia for example, although the rights expire after 20 years *from their granting* rather than the date of filing of the application, and for trees and vines they expire after 25 years rather than 20 years.

In Australia, the original Plant Variety Rights Act 1987 was superseded by the PBR Act 1994 conforming to the 1991 UPOV Act, and allowing all new cultivars of plants, including transgenic plants, fungi, and algae to be eligible for protection.

Protecting a plant using Plant Breeder's Rights

The PBR system in Australia is similar to other UPOV countries and will act as an example. Contact the PBR Office for a specific country for details pertaining to that country (see **Part 4 – Plant Name Resources**).

To use PBR in Australia you must apply to the PBR Office, which is now a department of Intellectural Property (IP) Australia in Canberra. The PBR Office confers exclusive commercial rights to cultivars (referred to confusingly as 'varieties') registered with the PBR Office. Cultivars protected by PBR may only be produced for sale, sold, imported, and exported with the authority of the owner. The Rights are a form of intellectual property, like patents and copyright.

Only new or recently commercially exploited cultivars can be registered.

- A new PBR 'variety' cannot be sold without the breeder's consent.
- A recently exploited variety is one which has been sold with the breeder's consent:
  - for up to 12 months in Australia for Australian-raised cultivars
  - for up to 4 years for overseas-raised cultivars sold in Australia (unless they are trees or vines, in which case they can have been sold for up to 6 years).

To be eligible for protection, the applicant must show that the new variety is distinct from all other cultivars of common knowledge, uniform and stable. In tree and vine PBR 'varieties', PBR continues for 25 years from the date of granting, and in all other PBR 'varieties', for 20 years from the date of granting. The new PBR 'variety' may be imported or produced in Australia.

Any plant registered under a statutory registration authority, such as the PBR Office in Australia, is validly published or 'accepted' according to the *Cultivated Plant Code*. The official publication of the Australian PBR Office is the *Plant Varieties Journal*, which contains general information as well as descriptions of newly registered cultivars. Until 2004, it was available in hard copy, but it is only available online now and can be downloaded directly from the internet: (www.ipaustralia.gov.au/pbr/journal_download.shtml).

Each new application for a name is recorded by UPOV and, being the first name chosen by the breeder, this name is also the true botanical name; it must therefore conform with the rules and recommendations of the *Cultivated Plant Code*.

COMMERCIAL SYNONYMS

In Australia, registrants are allowed the use of an alternative name to the scientific one, and this is referred to as the *commercial synonym*. A rose may, for instance, be imported under a difficult German name. The use of a commercial synonym assists marketing by avoiding the use of potentially unappealing UPOV names (which may be in a foreign language or consist of an unwieldy breeder's code). Unfortunately, this means that sometimes the same plant can be sold under more than one name, and since other countries in UPOV may choose different commercial synonyms, this further complicates communication about the plant.

PLANT BREEDER'S RIGHTS SYMBOLS

If a name is protected under PBR and is the true cultivar name (UPOV name) it is recommended, but not compulsory, that in publications and on plant labels the PBR logo (Figure 29) is placed after the cultivar name (which is written in single quotes). This indicates that the cultivar and its name are protected under law and a text warning about legal infringement may also be added. Note that it is not only the cultivar that is legally protected from propagation by other people but also the registered PBR name and its commercial synonym. Commercial synonyms are trade designations, not scientific names, and should not be written with the single quotes needed for true cultivars; this helps to distinguish the true scientific name from the commercial synonym.

Figure 29: The Plant Breeder's Rights symbol

PBR offices for various countries can be located through the International Union for the Protection of New Varieties of Plants (UPOV) website (www.upov.int/).

For some of the countries, see **Part 4 – Plant Breeder's Rights**.

Trademarks

A trademark is a letter, number, word, phrase, sound, smell, shape, logo, picture, aspect of packaging or combination of these used to distinguish the goods or services of one trader from those of others.

Trademarks are used as marketing tools so that the public can associate a certain quality and image with goods and services bearing the trademark, a procedure known as 'branding': they distinguish the goods or services of one trader from those of another. Coca Cola and Toyota are good examples of brand names. Trademarks cannot be descriptive of the goods or services, nor can they be geographical or a common surname. Well known common names cannot be used as trademarks because they are already established in the public domain and therefore said to be 'generic'.

At present, trademarks used for marketing plants are found mostly on the labels of annual bedding plants and roses.

In Australia, application for registration of trade-marks, like that of PBR, is from IP Australia and may take up to a year because of the checking required. A registration lasts for 10 years and then a renewal fee may be paid.

Plants fall into Class 31 of the International Classification of Goods and Services (plants, seeds, flowers and plant reproductive material). Related classes are 1, 5, 35 and 44. Trademark checking involves looking systematically at all words, pictures and symbols that are relevant to a trademark. Your trademark must be broken down into its 'device constituents'. For example, if your new trademark is a castle with a rose across it, then it would be necessary to look at all rose and at all castle pictures found in other trader's trademarks. The location of other trademarks containing these device constituents is found by searching the IP Australia trademarks database (www.ipaustralia. gov.au/trademarks/search_index.shtml). If your trademark involves more than words, then this process may take hours, even days, depending on the popularity of the trademark constituents that you have chosen. However, it is possible to pay a patent attorney to do this search for you.

In contrast to PBR, people are at liberty to propagate the cultivars with which a trademark is associated, provided there is no other legal protection; it is the trademark itself that is legally protected, not the plant. Naturally, brand names, which may be associated with one to many different kinds of plants, feature prominently on plant labels, but it is most important to realise that trademarks identify suppliers (manufacturers/producers). They do not identify products so they cannot, under law, be used as plant names: a brand name is not a plant name. The cultivar name of a plant may not be trademarked, nor may a trademark become the cultivar name of a new cultivar. The trademark may be used as a promotional name but it is not the true cultivar name of the plant. This legislation protects the originator from other people selling their plant under the same trademark, but there is nothing to stop other people from selling the plant under its real name or a different trademark.

Perhaps the single most confusing aspect of the names on plant labels at present is the confusion between trademarks and the true botanical names. It is important that customers recognise a trademark *as a trademark* on the plant label and not mistake it for a plant name. After all, if only a trademark appears on the front of a label it is not unreasonable to assume that the customer will both understand and use it as the plant name.

Trademarks may be registered or unregistered.

UNREGISTERED TRADEMARKS

These are essentially invented marketing names with some zing that are used to sell a plant; they have not been registered with the Trademarks Office of IP Australia or its equivalent in other countries. In law, they are known as unregistered (or common law) trademarks. Simply making up names like this may seem scandalous to some people, but it is well established that plants are frequently bought for their appealing names rather than their good looks. A rose by any other name might smell as sweet but a good promotional name might make it sell better! So, if you only have a long Latin botanical name to sell a plant, then this practice is hardly surprising, even if it does add to the general name confusion.

These names are not scientific names, common names or names that are registered for legal protection. However, it is possible, although extremely expensive, to defend them under common law rights. This is dependant upon demonstrating that the name has become established within the industry and that use, by another party, of the name is a misrepresentation which will deflect goodwill to the other party's business, causing damage to the business of the common law trademark owner. Legal defence of trademark infringements is much more straightforward if the trademark is registered.

Figure 30: Commonly used symbol for an unregistered trademark

A promotional campaign some 30 years ago sold *Ceanothus papillosus* var. *roweanus* under the name Blue Pacific. Incorrectly, *Ceanothus* 'Blue Pacific' can still be seen written as a cultivar name, although there is no such cultivar. Blue Pacific is merely a promotional name and has no validity under the *Cultivated Plant Code*.

REGISTERED TRADEMARKS

The owner of a registered trademark has the exclusive right to use (or authorise others to use) that trademark and to prevent other parties from using the same or a similar mark in respect of the products covered by the trademark registration (or in some cases, similar products). Registered trademark rights are usually much easier to enforce than common law rights. However, the exclusive right will be put at risk if the trademark is used by its owner as the name, or part of the name, of a cultivar. For labelling purposes, it is recommended practice for registered trademark owners to always include the botanical name of the plant on the label and position it well away from the trademark or marketing name. It should be added that trademarks only have legal force in their country of registration.

Registered trademarks may 'cover' or brand one or more products, so it is possible for a registered word or words to be uniquely linked with a particular product (plant). This is hardly different from providing the consumer with a name for the product, although, under law, such a trademark is definitely *not* a product name.

Figure 31: Commonly used symbol for a registered trademark

TRADEMARK SYMBOLS

Writing 'TM' next to a name does not confer any legal privileges whatever, although some people have taken it to mean that a registered trademark is pending (Figure 30). Statutory rights are granted only when a trademark is registered (the protection will then extend back to the date of application for that registration). There is no single official way of flagging that a trademark is registered. Once registered, the letter R in a circle, as ®, or the letters RTM are often used with the trademark, or there is an asterisk combined with text indicating trademark registration (Figure 31). However, it is not legally compulsory to give any warning at all of registration.

PROBLEMS CAUSED BY USING TRADEMARKS

Using trademarks is an understandable way of trying to improve sales. However, both registered and unregistered trademarks are used, wittingly or unwittingly, in the way that we usually use a common or botanical plant name; that is, as word(s) used to identify a particular object (rather than its manufacturer or supplier). This has the following consequences:

- it is hard keeping track of such names and ensure they are not mistakenly used as cultivar names.
- there is an increase in the number of names associated with particular plants, thus making it more difficult for people to know all the names

associated with a particular taxon, and to know that they are indeed exactly the same taxon, and
- there is a distancing of the consumer from the uniquely identifying name of the plant, its botanical name
- there is a proliferation of legally protected words associated with particular suppliers and therefore the increased potential for legal action.

In some ways, the biological principle of having only one name for each kind of organism is just as valid in the legal arena. Recent potential trademark infringements have highlighted legal problems over duplication of trademarks. Name databases help avoid duplication of legally protected names as well as scientific ones; however, the numbers of these legally protected names is increasing steadily with time, increasing the possibilities for litigation.

Fortunately, responsible nurseries usually incorporate the real name of the plant on the promotional label.

Mistaking trademarks for cultivar names
In Australia, before the Plant Variety Rights Act of 1987, trademarks were the most common method of legal protection used by breeders of new cultivars. However, now, under law, trademark names are separate and distinct from the true cultivar name. The existing cultivar name of a plant may not be trademarked, nor may a trademark become a cultivar name.

For a trademark to be accepted, it is important that its words do not conflict with other legally protected names or with existing botanical names. It is a condition of the 1996 Act that the words used for trademarks are not used as the name or part of the name of a plant. The IP Australia Office therefore checks trademark applications against names covered by PBR, names in nursery industry databases and other sources. This is because the trademark name effectively grants a legal monopoly over the words being used. For example, a number of Australian native plant cultivars had 'Australflora' as part of the name, such as Grevillea 'Australflora Canterbury Gold'. When the name 'Australflora' became a trademark, it was no longer possible to have this word as part of any cultivar name so these plants became Grevillea 'Canterbury Gold'.

The misleading use of a trademark in the form of a cultivar name is not permissible and may be legally challenged. Plant labels must clearly distinguish between the true botanical name and the trademark, and in the case of registered trademarks, they should also display the trademark application number.

Trademarking rights do not automatically carry over from one country to another. Confusion may arise when plants are introduced to Australia under the overseas trademark, which is then treated as a true cultivar name.

Relative benefits of trademarks and Plant Breeder's Rights

Trademarks and PBR names have different objectives. PBR protects a particular plant cultivar from being sold without the permission of the person who has registered their cultivar with the PBR Office; it also protects the name of the cultivar, preventing anyone from using the same name within that particular denomination class.

In contrast, a trademark identifies the goods of a particular trader. In other words, PBR gives legal protection over who may sell a particular plant cultivar, while for trademarks legal protection is only for the trademark itself; plants carrying a trademark may be distributed and sold by anyone, but not with the same trademark. PBR protects the cultivar and its name; trademarking gives exclusive rights to the trademarks themselves.

In lists of plants, there should be no need to include trademarks. However, where plants have become known under their associated trademark names (no doubt a deliberate intention of some traders) it may be necessary to include these together with the botanical name.

Trademarks and PBR names may be used together. In such a case, the trader would market the plant under its PBR name with the trademark also on the label. If a breeder has developed or imported a range or series of plants, then the trademark can be applied to these. This name can then be used beside the PBR name for each different cultivar in that series.

The Flower Carpet™ trademark for roses is an example of a trademark series:

- *Rosa* 'Noaschnee' ⑫ (Flower Carpet™ white)
- *Rosa* 'Noatraum' ⑫ (Flower Carpet™ pink).

In these examples, the UPOV names 'Noaschnee' and 'Noatraum' are, of course, also the botanical names and, as they are protected under PBR, they carry the PBR symbol ⑫, so that you know they are legally protected. It would have been possible when taking out PBR in Australia to have selected alternative commercial synonyms for these names. There are no universally agreed methods of presenting trademarked names. Here, the trademark Flower Carpet™ has been placed in brackets to separate it clearly from the other names, but there is no generally accepted convention.

Plant Breeder's Rights, patents and genetic engineering

Plants raised by genetic engineering techniques can be given cultivar names. The plants themselves cannot be patented, but in some cases the gene that has been introduced to produce the new cultivar can be patented. This patent cannot only produce royalties on sales but will also control the future use of the gene or new cultivar in other breeding programs.

For example, if a blue rose is produced by insertion of a gene that codes for a blue compound, the rose may be registered under PBR legislation with a cultivar name that conforms with the *Cultivated Plant Code*. The gene itself may be patented, and this patent will prevent other rose breeders using the gene (and the rose) in their breeding programs.

The company *Florigene* has pursued broad patent claim coverage in many countries around the world. These patents provide *Florigene* with legal protection in the company's specialist area, the manipulation of anthocyanin-based flower colour. Their patents include:

- PCT/AU92/00334 – Genetic sequences encoding flavonoid pathway enzymes and uses, therefore
- PCTAU96/00296 – Transgenic plants exhibiting altered flower color and methods for producing same.

The floral product branding affords recognition of *Florigene's* unique products and to complement *Florigene's* patent and trademark portfolios, PBRs have also been acquired.

As with PBR, patents are sometimes available for licensing; that is, the agreed use of the protected product by someone else, generally for a royalty.

using
Part three
plant names

Writing plant names

The ways of writing botanical names are not all specified in the two *Codes*, but there are, nevertheless, internationally accepted conventions for the presentation of plant names in printed text. Most notable is the use of italics for certain botanical names. This is almost universal, although occasionally underlining (a printer's symbol for italics) or some other typographical method may be used to differentiate names from the surrounding text.

In the examples below, the accepted conventions are given in blue. Examples in red are not recommended.

Family name

Botanical names above the rank of genus are sometimes written in plain type and sometimes in italics, but always with the first letter as a capital. We recommend the use of plain type:

Myrtaceae

Genus name

The genus or generic name is written with a capital first letter and italicised:

Banksia

Latin generic names cannot be placed into the plural by adding an 's' (as in English), or by changing the Latin formation of the word in any way.

'The bed was full of different *Banksias*.'
'The bed was full of different *Narcissi*.'

Both these are incorrect. It is better to say:

'The bed was full of different *Banksia*.'
'The bed was full of different *Narcissus*.'

Or, if this sounds strange, the names may be written as common names.

'The bed was full of different banksias.'

In the case of *Narcissus* (and *Gladiolus*), common usage suggests that 'narcissi' and 'gladioli' are acceptable, but these can only be written without italics as common names.

Specific epithet

This is written with a lower case first letter and italicised:

serrata

The use of a capital first letter was once the practice for some proper nouns, particularly names commemorating people, and it persists in some horticultural books such as *Hortus Third*. Thus *Eucalyptus muelleri* is recommended, not *Eucalyptus Muelleri*.

Species name

The specific epithet and species name are not the same thing. The specific epithet, when combined with a genus name, constitutes the name of a species:

Banksia serrata

If the species name is used repeatedly in a piece of writing, the genus name may be abbreviated to a capital letter and full stop (e.g. *B. serrata*) unless it begins a sentence or is used for the first time in a paragraph. There should also be no possibility of confusion with the names of other plants mentioned.

Remember that 'species' is written and pronounced the same way in both singular and plural: 'specie' is not a term used in botany.

Subspecies

The *Botanical Code* recommends that a subspecies be designated by the abbreviation 'subsp.' although 'ssp.' is also acceptable (but easily confused with the abbreviation spp., indicating more than one species. The abbreviation for 'subspecies' is written with a lower case first letter and is not italicised:

Eucalyptus globulus subsp. *pseudoglobulus*

Variety

Latin – *varietas*. Designated by the abbreviation 'var.', which has a lower case first letter and is not in italics:

Banksia spinulosa var. *cunninghamii*

Form

Latin – *forma*. Designated by the abbreviation 'f.', which is lower case and not in italics:

Cedrus atlantica f. *glauca*

Note: the rank designators subsp., var. and f. must always be used. Names without them are called undesignated trinomials and are to be avoided, even though they still occasionally appear in nursery catalogues and horticultural reference books, as they cause confusion over the exact rank of the taxon.

Cultivated variety (cultivar)

The cultivar name consists of at least a Latinised genus name (or unambiguous common name) followed by a cultivar epithet which, after 1959, must be in a language other than Latin.

The cultivar epithet is enclosed by single quotation marks and begins with a capital letter; it is not italicised or underlined. If there is more than one word, each additional word must start with a capital letter (except for words like 'of' or 'the', and words after a hyphen, unless they are proper nouns):

Camellia 'Donation'
Camellia 'Alba Plena'
Rhododendron 'Queen of Hearts'
Acer palmatum 'Shime-no-uchi'

The practice of omitting the quotation marks and, instead, placing 'cv.' before the cultivar name was abandoned with the 1995 *Cultivated Plant Code*:

Camellia cv. Donation
is incorrect

The cultivar name may follow either the botanical name of the species, or the name of the genus only, or the common name of the genus or species (provided that the common name is unambiguous). For instance, the Australian-raised golden pencil pine can be named:

Cupressus sempervirens 'Swane's Golden'
or *Cupressus* 'Swane's Golden'
or Pencil Pine 'Swane's Golden'

The generic name, followed by the cultivar name, is often used when the parentage of a particular cultivar is confused or it is impossible to link it with certainty to a particular species. Examples occur among the highly bred *Dahlia*, *Narcissus*, *Tulipa* and *Gladiolus*. An Australian example would be:

Grevillea 'Clearview David'

The conversion of botanical varieties to cultivar names in 1959 resulted in name changes. For example, *Chamaecyparis lawsoniana* var. *erecta* became *Chamaecyparis lawsoniana* 'Erecta' but not *Chamaecyparis lawsoniana erecta* as is sometimes written.

Hybrids

Hybrids between two species (interspecific hybrids) are indicated by inserting a hybrid sign (×) between the species names of the two parents. The names are written in alphabetical order, except that if the female parent is known then this is placed first:

Camellia japonica × *Camellia saluenensis*

The hybrid sign (×) should not be in italics and a single letter spacing should be left before and after it. Where the hybrid has been described according to the *Botanical Code* in the same way as a new species, the hybrid sign precedes the specific epithet without any spacing, thus:

Camellia ×williamsii

The printer's equivalent, the multiplication sign, is intermediate in size between the upper case and lower case '×' in a computer word program. If a multiplication sign is not available then a lower case '×' should be used, in which case a single letter space should be left on either side of it. Varieties or cultivars of hybrids are written thus:

Populus × *canadensis* var. *serotina*
Camellia × *williamsii* 'Donation'
Camellia 'Donation'
(*Camellia japonica* × *Camellia saluenensis*) 'Donation'

It is incorrect to precede a cultivar name with a hybrid sign to indicate its hybrid origin as in the following example:

Camellia × 'Donation'

Naturally occurring hybrids, if named at all, are named according to the *Botanical Code*. Most hybrids arising in the horticultural trade are simply given cultivar names.

Grevillea 'Poorinda Beauty', *Grevillea* 'Jeannie' and *Grevillea* 'Rachel' are all cultivars arising from crosses between *Grevillea alpina* and *Grevillea juniperina*. The absence of a specific epithet is a strong indication that the plant is of hybrid origin.

The cross between *Cupressus macrocarpa* and *Xanthocyparis nootkatensis* is an intergeneric hybrid that is known as Leyland Cypress; to show that it is an intergeneric hybrid, it has a hybrid sign placed in front of its botanical name without a space:

×*Cuprocyparis leylandii*

The generic (genus) name is formed from parts of the two parent generic names. Intergeneric hybrids, to be legitimate, must be described according to the rules of the *Botanical Code*. For example, any hybrid between a *Xanthocyparis* species and a *Cupressus* species has the generic name ×*Cuprocyparis*, but the specific epithet will differ depending on the parent species involved. For example, the cross *Cupressus glabra* × *Xanthocyparis nootkatensis* is ×*Cupressocyparis notabilis*. Any hybrid between *Cupressus macrocarpa* and *Xanthocyparis nootkatensis* is called ×*Cuprocyparis leylandii*, while any cross between *Cupressus glabra* and *Xanthocyparis nootkatensis* is called ×*Cuprocyparis notabilis*. A cultivar could be written in any of the following ways:

×*Cuprocyparis leylandii* 'Castlewellan Gold'
×*Cuprocyparis* 'Castlewellan Gold'
Leyland cypress 'Castlewellan Gold'

Group names

An assemblage of similar cultivars may also be designated a Group. Unlike a collective epithet, a Group does not necessarily include all the progeny of a particular cross and may contain similar cultivars of different parentage. The Group is not italicised.

Prunus Sato-zakura Group

If we wish to distinguish both the Group and a particular cultivar within that Group, then the Group name is written in brackets with the cultivar name following:

Prunus (Sato-zakura Group) 'Shirofugen'

Collective names and greges (grexes)

A collective name is a single designation covering all the offspring of a particular hybrid combination. The hybrid name *Camellia* ×*williamsii*, formed according to the *Botanical Code* and discussed above, is just one of these. A Group name is sometimes used this way; for example, the name *Iris* Dutch Group includes early flowering cultivars with a particular hybrid history. An individual cultivar would be written:

Iris (Dutch Group) 'Apollo'

Iris 'Apollo' if there was no wish to draw attention to the parentage of the cultivar. Another example of a collective name is *Rosa* Hybrid Tea.

For orchids, where the breeding history is often extremely complex, naming the different kinds can be extremely difficult, especially if you would like the name to indicate parentage. So in orchids, and only orchids, the word grex (plural greges, but often written grexes) is used instead of a Group. The Group name for the cross *Paphiopedilum* Atlantis grex × *Paphiopedilum* Lucifer grex is *Paphiopedilum* Sorel grex.

Graft-chimaeras

Graft-chimaeras (loosely known as graft hybrids) originate from a callus at the junction of stock and scion. They contain tissue from both stock and scion but are stable and can be propagated vegetatively. They are named under the *Cultivated Plant Code* and may be written as a formula which is the names of the taxa in alphabetical order linked by a '+' with a space on either side thus:

Cytisus purpureus + *Laburnum anagyroides*

This graft-chimaera may be given a new 'genus' name by linking part of the name of one genus with part of that of the other, connected by a vowel. The '+' is then placed at the front without a space between it and the name of the new genus.

+Laburnocytisus

The name of a cultivar of such a graft-chimaera might then be:

+Laburnocytisus 'Adamii'

The most commonly produced graft-chimaeras are in *Camellia*: they are usually simply given cultivar names.

Synonyms

Synonyms are outdated or 'alternative' names. Unfamiliar new names are best accompanied by their old names, so that people are not confused by the new name. The synonym may be strictly a prior botanical name or, occasionally, a name of no botanical standing or a frequently used misspelling. The usual way of doing this is to put the old name in brackets with 'syn.' in front of it, although the 'syn.' is sometimes omitted.

Lophostemon confertus (syn. *Tristania conferta*)

Uncertain names

When there is an element of doubt about the identification of a plant, a question mark is put in front of the full name:

?Davidia involucrata

The meaning is: this is perhaps *Davidia involucrata*.

When the genus is known but there is some uncertainty about the species, a question mark is put in front of the specific epithet:

Pinus ?aristata

The meaning is: this is a *Pinus* species, possibly *P. aristata*.

When the plant may be an extreme variant or hybrid, the abbreviation 'aff.' is put before the specific epithet (aff. = affinis, having affinity with) as in:

Tilia aff. *americana*

This means that the plant is very close to *Tilia americana* but does not agree sufficiently with descriptions to allow definitive identification.

Sometimes 'sp. aff.' is used to indicate a variant that could become a new species in the future as in:

Thelymitra sp. aff. *megcalyptra* (Alpine)
Thelymitra sp. aff. *megcalyptra* (North-west)

If a plant is very similar to another species (but clearly not the same) then 'cf.' may be used:

Tilia cf. *americana*

It differs from 'aff.' by not implying any relationship with the denoted species. If only the genus is known, then it is conventional to write the abbreviation 'sp.' for the specific epithet:

Callistemon sp.

The plural of 'sp.' is 'spp.', thus:

Melaleuca spp.

If the cultivar is not known, then the abbreviation 'cv.' is used (plural 'cvs'):

Camellia japonica cv.
or *Camellia* cv.

If a plant has not yet been described, it may be designated as follows:

Banksia sp. (undescribed)

This means that all details are known about the plant but it has not yet been described according to the rules of the *Botanical Code*. This is sometimes written *Banksia sp. nov.* (*species nova* = new species), a convention that should only be used in the publication in which the species is first described, the *sp. nov.*, in italics, following the new epithet.

Coffea bridsoniae A.P.Davis & Mvungi *sp. nov.*

Similarly, when a specific epithet is transferred to a new genus, as when *Tristania conferta* became *Lophostemon confertus*, this is known as a new combination and appears in the publication where the combination is made as:

Lophostemon confertus (R. Br.) Paul G. Wilson & Waterhouse *comb. nov.*

Common names

There is no universally accepted way of writing common names. However, the following is generally recommended:

- For a name used in a general sense covering a group or genus (e.g. bottlebrush, conifer, oak) start with a lower case letter; this also applies to botanical names used in a general sense (e.g. banksias, camellias and acacias).
- If one particular species or plant is referred to then we suggest that you use capitals for the first letter of all words, except when there is a hyphen between two words:

River Red Gum
Lemon-scented Gum

Some publishers use lower case letters for all common names. This can lead to confusion as to what exactly constitutes the common name in a piece of text. What, for instance, is the common name in the sentence: 'In the centre of the garden was a red flowering gum'?

Do not put English or vernacular names in either single or double quotation marks, as these may be confused with the single quotation marks used to designate a cultivar name. Of course, double quotation marks should never be used for cultivar names.

Hyphens

In general, hyphens are not permitted in compound names, although there are a few exceptions where the two words are not normally combined; for example, the nouns *Coix lacryma-jobi* (Job's Tears) and *Aster novae-angliae* (New England). The *Cultivated Plant Code* does not specify how hyphens are to be used in cultivar names but it would appear sensible to follow the example of the *Botanical Code*:

Sparrmannia africana 'Alboplena'
(**not** 'Albo-plena' or 'Albo-Plena')

The *Cultivated Plant Code* cites 'Go-go Dancer' as an acceptable name, but cautions against simply applying hyphens in a way that might seem logical and cites the case of 'Terra-Cotta' being corrected to 'Terracotta', as the name was coined in reference to the colour not the proper name.

Spelling

Occasionally there is disagreement, even among botanists, over the correct spelling of a plant name. You may notice this when reputable publications use slightly different spellings for the same name. For example, is it *Brachyscome* or *Brachycome*, *Buddleia* or *Buddleja*, *Lechenaultia* or *Leschenaultia*, *Wistaria* or *Wisteria*?

Botanists, like other people, are irritated by apparent spelling inconsistencies in plant names, but they are not at liberty to choose what might seem the most appropriate spellings. Disputed cases like those quoted are relatively rare, and to resolve them, an appeal is generally made to a particular interpretation of the *Code*. Unfortunately interpretations sometimes differ and universal agreement is not always obtained. To resolve the issue, a proposal recommending the conservation of a particular spelling may be submitted to a special committee appointed by the International Association of Plant Taxonomists.

In most cases, the spelling of a name is unambiguous according to the *Code*. For instance, it might be thought that a genus named after Prof.

Johann G. Gleditsch, an 18th century German botanist, should be called *Gleditschia*. The spelling *Gleditsia* was, however, a deliberate latinisation by Linnaeus; to spell it otherwise is to contravene the *Code*.

Important items of the Code that must be taken into account in these cases are:

- the recommendation that the spelling of a name used by the person who first applied it should be adopted
- that it should be retained except for obvious printing or spelling errors
- that the liberty of correcting a name should be used with reserve, and
- that the formation of Latinised words should follow classical usage as far as possible.

Disagreement over the formation of some specific epithets has led the *Botanical Code* to prescribe rules which have resulted in irritating minor changes. To illustrate these, the formation of specific epithets must be briefly introduced. All hyphens used in the discussion on the formation of epithets are purely to reveal word structure.

The specific epithet may be a noun or an adjective. When it is a noun, it is usually in the genitive (possessive) case. In *Cedrus libani*, libani means 'of Lebanon' and is the genitive case of *libanus*, Lebanon.

When such epithets are derived from the names of people, they are formed by adding -*i*- and an appropriate genitive ending to the name. Therefore, *Banks-i-i* means of (Sir Joseph) Banks, and *banks-i-ae* means of (Lady) Banks. When the name ends in a vowel, -*y* or -*er*, the linking -*i*- is omitted.

guilfoylei	Guilfoyle's
ashbyae	(Ms) Ashby's
muelleri	Mueller's

The names of people may be converted into adjectives by adding -*i*- with -*anus*, with exceptions as before. Thus:

banks-i-anus	in honour of Banks (or Lady Banks)
guilfoyle-anus	in honour of Guilfoyle
ashby-anus	in honour of Ashby (or Mr Ashby)

but not *mueller-anus* (compared with *muelleri*) but *mueller-i-anus*. The *Botanical Code* has recently reverted to the form *mueller-i-anus* after standardising briefly on *muelleranus*.

The linking letter -*i*- has also caused confusion in a number of compound epithets like *salviifolius* (sage-leaved) and *podalyriifolia* (*Podalyria*-leaved).

When such compounds are formed, the *Code* prescribes linking the stem of the first word to the second word with *-i-*. Thus:

Latus	*lat*	+	*i*	+	*folia*	=	*latifolia*	(broad-leaved)
Rosa	*ros*	+	*i*	+	*folia*	=	*rosifolia*	(rose-leaved)
Salvia	*salvi*	+	*i*	+	*folia*	=	*salviifolia*	(sage-leaved)
Ficus	*fic*	+	*i*	+	*folia*	=	*ficifolia*	(fig-leaved)

An alternative way of forming the compounds using the genitive, viz. *rosae* + *folia* or *salviae* + *folia,* is not accepted by the *Code.* Epithets such as *podalyriaefolia* and *rosaefolia* should be corrected to *podalyriifolia* and *rosifolia.*

If you are not sure how to spell a name, then you can consult one of the texts listed in **Part 4 – Plant name resources** or contact your nearest herbarium or botanic garden.

The structure of Latin names

The name of a genus is a noun which in Latin will be masculine, feminine or neuter in gender. Most trees and shrubs are feminine except for those whose names are based on Greek words.

The specific epithet can be either an adjective or a noun. If it is an adjective, it has the same gender as the name of the genus and it generally either describes a prominent feature of the plant, or commemorates a person (often a botanist, explorer or plant collector) or place. Examples of adjectival epithets are: *prostratus,* prostrate; *pauciflora,* few-flowered; *hookerianus,* in honour of Hooker; *sylvaticus,* belonging to the woods; *capense,* from the Cape.

Although there are rules for the appropriate endings of adjectival specific epithets, they are too diverse to consider here. Those who wish to go further should consult *Botanical Latin* (Stearn 1992). A few examples that regularly cause difficulties in spelling have been considered in the previous chapter.

Pronunciation

People often worry that they might be showing their ignorance if they cannot pronounce botanical names fluently and in the 'proper' way, but pronunciation is not of vital concern provided others can understand what is being said.

Which Latin do we use?

There is no general agreement on how Latin should be pronounced. In English-speaking countries there are two main systems: the traditional system which botanists and gardeners use, and the reformed academic pronunciation which aims to reconstruct the spoken language of classical Rome. This latter is closer to the pronunciation of plant names in many European countries, although people proficient in one of these tend to pronounce words by analogy with their own language. We recommend, however, the traditional English system as indicated below.

Part 4 – Plant name resources gives references to assist with pronunciation.

General guidelines

If you wish to take more trouble over pronunciation then the following general guidelines may help you.

STRESS ON SYLLABLES
- In Latin every vowel is usually pronounced; for example, co-to-ne-as-ter not cot-on-easter as is our inclination. So, when first striking a new

and tricky botanical name, try pronouncing each syllable at a time: *Hebe* has two syllables, *Nerine* has three and *Cordyline* four.

- Accentuation of botanical names also depends on the Latin or Greek roots. Many names are wrongly pronounced for this reason and there are many names like *Callistemon* and *Phyllocladus* that few people accent correctly. In our opinion, it is more important that the names be communicated than that they should be pronounced correctly.

In general, words of two syllables have the stress on the first, as in RO-sa. Words of three syllables have the stress placed on the penultimate syllable if that syllable has a long vowel (e.g. Au-CU-ba) or on the antepenultimate (third last) syllable if the penultimate syllable is short (e.g. FLA-vi-dus). In words of four syllables, the accentuation is often wrongly placed, even by botanists. CALL-i-STE-mon is correct, whereas it is generally pronounced Call-IS-te-mon. EU-ca-LYP-tus is correct as also is Chry-SAN-the-mum. The rule is the same as for three syllables.

Several books referred to in **Part 4 – Plant name resources** list common endings of words and rules for their accentuation.

SHORT AND LONG VOWELS

The list on the opposite page gives the traditional English horticultural pronunciation of Latin vowels and consonants. Some local variants are included. Without a knowledge of the Latin or Greek roots on which these words are based, it is impossible to know which vowels are short or long.

PEOPLE AND PLACES

Many botanical names, especially those of genera, commemorate a person or place; for example, *Clivia, Camellia, Fuchsia, Choisya, Degeneria, Vouacapoua, Warszewiczia* and *Syreitschikovia*. Some of these may be difficult to pronounce or, like *Eschscholtzia*, difficult to spell. There are three options for pronouncing these names: as if they were Latin, as if they were English, or as they would be pronounced in the original language. *Choisya*, named after the Frenchman Choisy, is usually pronounced as if it were English, CHOI-sya, not choi-SY-a as in French or KO-I-S-YA as in Latin. *Quillaja*, pronounced as in Spanish, is barely intelligible to English-speaking people. Even with three options, common usage may choose a fourth; for example, the usual pronounciation of *Camellia* follows neither Latin, English or German.

There is no rule; the best solution is to use the commonly-accepted pronunciation or the one that is most pleasing to the ear.

GUIDE TO PRONUNCIATION OF BOTANICAL NAMES

a   Short as in cat; long as in rather; not as in gate

ae   As in seat

au   As aw in shawl

c   Before a, o, u as in cat; before e, i, y as in ceiling

ch   (Greek words) K as in car or ch as in chair (depending on context)

e   Short as in let; long as in meet

ei   As in height

g   Before a, o, u as in gone; before e, i, y as in gem

I   Short as in tin; long as in fine

J   As in jet

o   Short as in pot; long as in vote

oe   As in see

ph   As f

s   As in this, not those

u   As in rub

ui   As in ruin

v   As in vet

y   Short as in Robyn; long as in gyrate

Remembering
names

The *RHS Plant Finder* for the years 2004–2005 (Lord *et al.* 2004) lists over 73 000 different plants available in nurseries in the UK, while the Aussie Plant Finder of 2004 lists about 35 000 in Australian nurseries. How can we begin to remember even a small proportion of these names?

READING

Naturally, the more you use plant names the sooner their use will become second nature. By reading gardening books and visiting nurseries where you study the plant labels and catalogues, you will soon become familiar with the commonly grown plants. It is generally the first sight of a botanical name that worries people. We all cope well with the familiar *Chrysanthemum*, which is a word with quite a complicated construction, but feel unsure when confronted with a name like *Chionodoxa* because it is unfamilar.

PRONUNCIATION

The ability to pronounce a name will help you to remember it.

WORD DERIVATIONS

If you really want to master plant names then try and learn the meanings of commonly used Latin and Greek words and the parts from which they are constructed. A good etymological dictionary will have words with the same roots as many plant names and will give the Latin origin; for example, prostrate from *prostratus*; foliage from *folia*. Some dictionaries (e.g. the *Chambers 20th Century Dictionary*; Geddie 1959) contain many botanical terms which are Anglicisations of Latin descriptive words often used as specific epithets; for example, *rostrate*, beaked; *angustifoliate*, narrow-leaved. A little time spent learning a few Latin words will pay enormous dividends, not only in helping to remember the names but also in learning much more about the plants themselves. You have probably seen the specific epithet *sempervirens*. This can be translated as follows: *semper* = always, *virens* = green. Together they mean 'evergreen', as in *Buxus sempervirens*, Box, or *Cupressus sempervirens*, Italian Cypress. Before long, you will be able to treat many names as small descriptions that will help you to remember them, although not all specific epithets are descriptive and some may be puzzlingly inappropriate for the plants.

A number of useful books to help you in translating Latin names are listed in **Part 4 – Plant name resources.**

Recommended format for nursery plant labels

Plant labels in retail nurseries now carry a plethora of names of uncertain botanical and legal status that are as confusing to industry employees as they are to the general public. This is not altogether surprising as a plant label in a nursery is expected to satisfy all the following people and organisations: the retailer; botanists; the PBR Office; the label-producing company; patent and trademark attorneys; and, last but not least, the customer.

It is clearly in the general interest to ensure that litigation based on legally protected names is reduced to a minimum. It would also seem desirable that the one universal and uniquely identifying name, the scientific name, which is generally the key to further information about the plant, should appear on each plant label.

One major way to assist all label-users would be to clearly identify, on the labels, the legal and scientific status of the names that are being used. This can be done by presenting each kind of name in a distinctive way (Figure 32).

We have identified eight different kinds of names that now appear quite frequently on retail plant labels. Using the words 'spring splendour', we have shown in Table 5 how the particular kind of name can be indicated by the way these words are displayed on the label.

All reasonable steps must be taken to ensure labels are accurate. Occasionally labels may be misleading or deceptive by creating a false impression or making false claims, and obviously this is to be avoided.

Razzle Dazzle®

Anigozanthos manglesii ' Little Jumper'

A dwarf red and green Kangaroo Paw

~~~~~~~``````~~~~~`~`~~~`~~~~`~~~`~
~~~~~`~~~``````~~````````~~~~`~`~`~~`~
~`~``~~~~``~~~~~```````~~`~~```~~~``~~~
`~~``~~~``~~~~~```~~~~`~~~~``~~~`~~`~
~`~`~~~~`~``````~~``~~`~~~~`~

USES

~~`~`````~~`~`~~`~~~~``~~~`~`~~~~`~`~~
~~`~~~~`~~`~~~~~~``````~``````````~~~~``
`~~~`~`````~`~``~`~`~~`~~`~~~~`

CULTURE

~`~~~`~~~`~~~`~~~~~`~~~~`~~~`~`~~~~
`~~~~`~~~~~```````~`~~`~~~~~~~`~`~

Produced under licence by Big Red Nurseries, Mumbilla.
The name RAZZLE DAZZLE is protected by trademark.

Anigozanthos manglesii 'Little Jumper'

Figure 32: A mockup label illustrating the various names used on retail nursery labels. Suggested names to incorporate on labels are the botanical name, including the cultivar or Plant Breeder's Rights name or its commercial synonym.

Table 5: Recommendations for formatting commercial nursery labels. This is a guide only and not a substitute for legal advice.

Type of name	Format of name	Comments
Botanical name	*Grevillea rosmarinifolia*	The botanical name is the single unique identifier for the plant and should be placed somewhere on the label. It may be put on the back of the label when the front is used for strong promotion. Botanically, this is the species name consisting of the genus and specific epithet. If the plant is a cultivar of this species it would be written as in the next example.
Cultivar name	*Grevillea rosmarinifolia* 'Spring Splendour'	In this botanical name, the words 'Spring Splendour' in single quotes are known botanically as the cultivar epithet and this kind of botanical name is often referred to as the cultivar name. As presented here, the cultivar has no legal protection.
Synonym	*Corymbia citriodora* (syn. *Eucalyptus citriodora*)	Alternative or old names are placed in brackets after the botanical name. The synonym is placed immediately after or under the botanical name.
Unregistered common law trademark	SPRING SPLENDOUR™ *Grevillea rosmarinifolia*	Trademarks are generally placed on the front of labels as promotional brand names.
Registered trademark	SPRING SPLENDOUR® *Grevillea rosmarinifolia*	Note: • the trademark cannot be used as the botanical or cultivar name of the plant or as a substitute for those names • There are no absolute rules on writing trademarks. However, in general a plant trader's trademark is given an asterisk or the letters ™ are written beside it when it is found on packaging and advertising. The symbol ™ is generally taken to indicate a pending registration, while the symbols ® or RTM indicate a registered trademark with full legal protection. We recommend this usage even though it has no legal force. • It is recommended that the trademark be written in capital letters or possibly a fancy script or bold colour that is different from the botanical or cultivar names.

Type of name	Format of name	Comments
		• A particular trademark should be used consistently in the same way on all labels.
		• Somewhere on the label the trademark should be followed by the botanical and/or cultivar and/or common name, for example:
		EVERGREEN EDGER® *Buxus sempervirens* 'Rotundifolia', Round-leaf Box.
		• Where a trademark is used under licence from another party it should be used in accordance with the licence agreement and it is recommended that licensing be indicated on the label, for example:
		EVERGREEN EDGER® *Buxus sempervirens* 'Rotundifolia', Round-leaf Box, and is used under licence.
		• Sometimes a copyright notice may appear on the label to protect the artistic material or photographs appearing on the label, for example:
		© Copyright 2005 – Gobsmackers Nursery
A true cultivar name protected by PBR	*Grevillea rosmarinifolia* 'Spring Splendour' ⓟ	All names protected under PBR legislation would carry the PBR symbol or PBR letters.
A commercial synonym protected by PBR	*Grevillea rosmarinifolia* Spring Splendour ⓟ	Under PBR legislation a commercial synonym can be used if the true cultivar name is not appropriate for marketing; for example, a breeder's code may be the true cultivar name. The commercial synonym is not written with single quote marks.
A plant protected by PBR and propagated under licence	*Grevillea rosmarinifolia* 'Spring Splendour' ⓟ is used under licence	Where a PBR-protected plant is propagated and sold under licence from another party, it should be used in accordance with the licence agreement and it is recommended that licensing be indicated on the label.
Common name	Spring Splendour	Common names are "generic" and therefore cannot be used as trademarks. They are written without quotes or any other embellishment or symbol.

plant

Part four

name resources

Books and websites to help with plant names

Accurate lists of botanical names

Botanical taxonomic research is published in scientific books, papers and journals; these are the primary source of name changes that are part of ongoing botanical research. Unfortunately, these sources are not readily available to the general public but are housed in research institutes such as herbaria and university libraries. Usually the keen horticulturist must depend on botanically checked and reliable secondary sources, and if these are published in hardcopy form they inevitably become outdated. There are now many plant lists and floras available through the internet, and often these are regularly updated. If there is a difficult name that you cannot sort out, then contact your nearest botanic garden.

Families

Heywood VH (Ed.) (1993). *Flowering plants of the world.* BT Batsford Ltd, London.

Morley BD and Toelken HR (1983). *Flowering plants in Australia.* Rigby Publishers, Adelaide.

Genera

Farr ER, Leussink JA and Stafleu FA (Eds) (1979). *Index nominum genericum (plantarum). Volumes 1–3.* Bohn, Scheltma and Holkema, Utrecht.

Farr ER, Leussink JA and Zijlstra G (Eds) (1986). *Index nominum genericorum (plantarum). Supplement I.* Bohn, Scheltema & Holkema, Utrecht.

Mabberley DJ (1997). *The plant-book.* 2nd edn. Cambridge University Press, Cambridge. (A listing of current generic names placed in the families of the Arthur Cronquist system.)

Willis JC (1973). *A dictionary of the flowering plants and ferns.* 8th edn. Revised by HK Airy Shaw. Cambridge University Press, Cambridge. (A comprehensive listing for its time of all genera, whether in current use or not; however, it places the genera in a family system that has now become rather outdated.)

Lists of validly published names, not necessarily current

Chapman AD (1991). *Australian plant name index.* Australian Government Publishing Service, Canberra. (This is a listing of all names published for Australian flora and where they were published, whether they are current or a part of the taxonomic history of Australia. The listing is now available online at www.anbg.gov.au/cpbr/databases/apni.html)

International Plant Names Index, www.ipni.org/index.html (This is a compilation of Index Kewensis of seed plants from the Royal Botanic Gardens Kew, the Gray Herbarium Index of Harvard University of vascular plants of the New World and the Australian Plant Name Index of Australian taxa. Since 2004 it has included Index Filicum, that lists names of ferns.)

Royal Botanic Gardens Kew (1997). *Index Kewensis.* On compact disc. Oxford University Press, Oxford. (This is a listing of the places of publication for all new species of flowering plants. Amounting to a list of the names of all known flowering plants, it is one way in which professional botanists can quickly find out where the original descriptions of particular species may be found, and is also a way of keeping abreast of recent additions and changes. Supplements to *Index Kewensis* are produced every 5 years.)

Floras and checklists of currently accepted plant names

The following is not a complete list, there being a number of additional resources published but not widely available.

AUSTRALIA

Australia Biological Resources Study. *Flora of Australia online*, www.deh. gov.au/biodiversity/abrs/online-resources/flora/main/index.html

Buchanan AM (Ed.) (1999). *A census of the vascular plants of Tasmania and index to the student's flora of Tasmania*. 3rd edn. Tasmanian Herbarium Occasional Publication No. 6. Tasmanian Herbarium, Hobart.

George A (Ed.) (1980–). *Flora of Australia*. Australian Biological Resources Study, Canberra. (Later edited by others, and later published by the CSIRO, Melbourne).

George Brown Darwin Botanic Gardens (2004). *Parks and Wildlife Commission of the NT Herbarium Database Species Checklist*, www. nt.gov.au/ipe/pwcnt/index.cfm?attributes.fuseaction=open_page&page_ id=7352

Henderson RJF (ed.) (2002). *Names and distribution of Queensland plants, algae and lichens*. Queensland Herbarium, Toowong, Queensland.

Hnatiuk RJ (1990). *Census of Australian vascular plants*. Australian Flora and Fauna Series Number 11. Australian Government Publishing Service, Canberra.

National Herbarium of New South Wales (1999–2005). *NSW plants search site* (plantnet.rbgsyd.nsw.gov.au/search/florasearch.htm) Royal Botanic Gardens & Domain Trust, Sydney.

Plant Biodiversity Centre. *The electronic flora of South Australia* (flora.sa. gov.au/). Plant Biodiversity Centre, Adelaide.

Ross JH and Walsh NG (2003). *A census of the vascular plants of Victoria*. 7th edn. Royal Botanic Gardens, Melbourne. Also available on the internet at www.rbg.vic.gov.au/plant_science/online_plant_information/ viclist (The *Census of the Vascular Plants of Victoria* is a list, compiled by botanists at the National Herbarium of Victoria, of the names of all the native and naturalised plants occurring in Victoria. It can be taken as a list of recommended names; they have been carefully examined and assessed and do not necessarily include those most recently published if the taxonomic decision is considered unwarranted.)

Western Australian Herbarium. *FloraBase* (www.naturebase.net/content/ view/2452/1322/) Department of Conservation and Land Management, Perth.

PACIFIC

Allan Herbarium (2000). *New Zealand Plant Names Database*. Landcare Research New Zealand, Lincoln (nzflora.landcareresearch.co.nz/).

Papua New Guinea National Herbarium (LAE) *Plants of Papua New Guinea* (www.pngplants.org) Forest Research Institute, LAE, Papua New Guinea, a division of the Papua New Guinea Forest Authority.

Parsons MJ, Douglass P and Macmillan BH (1995). *Current names list for wild gymnosperms, dicotyledons and monocotyledons (except grasses) in New Zealand as used in Herbarium CHR.* Landcare Research New Zealand, Lincoln.

Asia

Czerepanov SK (1995). *Vascular plants of Russia and adjacent states (the former USSR).* Cambridge University Press, Cambridge.

Flora of China Editorial Committee (1994–). *Flora of China.* Science Press, Beijing, and Missouri Botanical Garden Press, St Louis. (Accessible on the internet at flora.huh.harvard.edu/china/)

Iwatsuki K, Yamazaki T, Boufford DE and Ohba H (Eds) (1993–) *Flora of Japan.* Kodansha Ltd, Tokyo.

Japanese Society for Plant Systematist. *Flora of Japan,* foj.c.u-tokyo.ac.jp/gbif/

Missouri Botanical Garden. *Flora of China Checklist* (mobot.mobot.org/W3T/Search/foc.html) Missouri Botanical Garden, St Louis.

New York Botanical Garden. *Flora of the Caucasus Region,* sciweb.nybg.org/science2/hcol/geor/index.asp

Turner IM (1995). A catalogue of the vascular plants of Malaya. *The Gardens' Bulletin Singapore* 47, 1–757.

Europe

Royal Botanic Garden Edinburgh. *Flora Europaea,* 193.62.154.38/FE/fe.html

North and South America

Balick MJ, Nee MH and Atha DE (2000). *Checklist of the vascular plants of Belize.* Memoirs of the New York Botanical Garden volume 85. The New York Botanical Garden Press, New York.

Brako L and Zarucchi JL (1993). *Catalogue of the flowering plants and gymnosperms of Peru.* Missouri Botanical Garden, St. Louis.

Conservatoire et Jardin botanique de la ville de Genève and Missouri Botanical Garden. *El Sistema de información Botánico Flora del Paraguay,* www.ville-ge.ch/cjb/bd/fdp/index.html

Harvard University Herbaria. *The Gray Herbarium Index of New World Plant Names* (zutto.huh.harvard.edu/databases/) Harvard University Herbaria, Cambridge.

Instituto de Botánica Darwinion. *Catálogo de las Plantas Vasculares de la República Argentina. II.* www.darwin.edu.ar/Publicaciones/ CatalogoVascII/CatalogoVascII.asp

Jørgensen PM and León-Yánez S (1999). *Catalogue of the vascular plants of Ecuador.* Missouri Botanical Garden Press, St. Louis. (Accessible on the internet at mobot.mobot.org/W3T/Search/cvpe.html)

Kartesz JT (1994). *A synonymized checklist of the vascular flora of the United States, Canada, and Greenland.* 2nd edn. Timber Press, Portland.

Kartesz J and Biota of North America Program. *A Synonymized Checklist of the Vascular Flora of the United States, Puerto Rico, and the Virgin Islands* (www.csdl.tamu.edu/FLORA/b98/check98.htm) North Carolina Botanical Garden, Chapel Hill.

Maticorena C and Quezada M (1985). Catálogo de la flora vascular de Chile. *Gayana Botanica* 42, 1–157.

Missouri Botanical Garden.W3TROPICOS (mobot.mobot.org/W3T/ Search/vast.html) Missouri Botanical Garden, St Louis.

Missouri Botanical Garden, Instituto de Biologia of the National Autonomous University of Mexico and the Natural History Museum, London. *Flora Mesoamericana,* www.mobot.org/MOBOT/FM/intro.html

Seymour FC (1980). *A check list of the vascular plants of Nicaragua. Phytologia Memoirs I.* Harold N Moldenke & Alma L Moldenk, Plainfield, New Jersey.

Zuloaga FO (1996). *Catálogo de las plantas vasculares de la República Argentina I, Pteridophyta, Gymnospermae y Angiospermae (Monocotyledoneae).* Missouri Botanical Garden, St. Louis.

AFRICA

Brochmann C, Rustan OH, Lobin W and Kilian N (1997). *The endemic vascular plants of the Cape Verde Islands, W Africa. Sommerfeltia 24.* Botanical Garden and Museum, University of Oslo, Oslo.

Calane de Silva M, Izidine S and Amude AB (2004). *A preliminary checklist of the vascular plants of Mozambique. Southern African Botanical Diversity Network Report No. 30.* SABONET, Pretoria.

Germishuizen G and Meyer NL (Eds) (2003). *Plants of southern Africa: an annotated checklist. Strelitzia 14.* National Botanical Institute, Pretoria.

Hansen A and Sunding P (1985). *Flora of Macaronesia. Checklist of vascular plants. 3 revised edition. Sommerfeltia 1.* Botanical Garden and Museum, University of Oslo.

Kobisi, K (2005) *Preliminary checklist of the plants of Lesotho. Southern African Botanical Diversity Network Report No. 34.* SABONET, Pretoria, and Roma, Lesotho.

Mapaura A and Timberlake J (Eds) (2004). *A checklist of Zimbabwean vascular plants. Southern African Botanical Diversity Network Report No. 33.* SABONET, Pretoria, and Harare.

Phiri PSM (2005). *A checklist of Zambian vascular plants. Southern African Botanical Diversity Network Report No. 32.* SABONET, Pretoria.

HORTICULTURAL FLORAS AND CHECKLISTS

Cullen J (1995). *The European garden flora. Volume IV. Dicotyledons (Part II).* Cambridge University Press, Cambridge.

Cullen J (1997) *The European garden flora. Volume V. Dicotyledons (Part III).* Cambridge University Press, Cambridge.

Cullen J (2000). *The European garden flora. Volume VI. Dicotyledons (Part IV).* Cambridge University Press: Cambridge.

Food and Agriculture Organization of the United Nations. *Hortivar – Horticultural cultivars performance database*, www.fao.org/hortivar/

Hibbert M (2004). *The Aussie plant finder.* Florilegium, Glebe, New South Wales.

Philip C (2005). *RHS Plant Finder 2005–2006.* Dorling Kindersley, London.

Royal Horticultural Society (2005). (www.rhs.org.uk/databases/summary. asp) Royal Horticultural Society, Wisley, UK.

Spencer RD (1995). *Horticultural flora of south-eastern Australia. Volume 1. Ferns, conifers & their allies.* University of New South Wales Press, Sydney.

Spencer RD (1997). *Horticultural flora of south-eastern Australia. Volume 2. Flowering plants, dicotyledons Part 1.* University of New South Wales Press: Sydney.

Spencer RD (2002). *Horticultural flora of south-eastern Australia. Volume 3. Flowering plants, dicotyledons Part 2.* University of New South Wales Press, Sydney.

Spencer RD (2002). *Horticultural flora of south-eastern Australia. Volume 4. Flowering plants, dicotyledons Part 3.* University of New South Wales Press, Sydney.

Spencer RD (2005). *Horticultural flora of south-eastern Australia. Volume 5. Flowering plants, monotyledons.* University of New South Wales Press, Sydney.

Staples, GW and Herbst, DR (2005). *A tropical garden flora*. Bishop Museum Press, Honolulu.

Trehane P (1989). *Index Hortensis. Volume 1: Perennials*. Quarterjack Publishing, Wimborne, UK.

Tucker AO, Kunst SG, Vrugtman F and Hatch LC (1994–1995). A sourcebook of cultivar names. *Arnoldia* 54, 1–59. (Accessible on the internet at herbarium.desu.edu/Cultivar_chklst.pdf)

United States Department of Agriculture, Agricultural Research Service, National Genetic Resources Program. *Germplasm Resources Information Network – GRIN Taxonomy Online Database*, www.ars-grin.gov/cgi-bin/npgs/html/index.pl

Walters SM, Brady A, Brickell CD, Cullen J, Green PS, Lewis J, Matthews VA, Webb DA, Yeo PF and Alexander JCM (Eds) (1986). *The European garden flora. Volume I. Pteridophyta, Gymnospermae, Angiospermae – monocotyledons (Part I)*. Cambridge University Press, Cambridge.

Walters SM, Brady A, Brickell CD, Cullen J, Green PS, Lewis J, Matthews VA, Webb DA, Yeo PF and Alexander JCM (Eds) (1984). *The European garden flora. Volume II. Monocotyledons (Part II)*. Cambridge University Press, Cambridge.

Walters SM, Alexander JCM, Brady A, Brickell CD, Cullen J, Green PS, Heywood VH, Matthews VA, Robson NKB, Yeo PF and Knees SG (Eds) (1989). *The European garden flora. Volume III. Dicotyledons (Part I)*. Cambridge University Press, Cambridge.

INTERNATIONAL CULTIVAR REGISTRATION AUTHORITIES

ICRAs are listed on the International Society for Horticultural Science website (www.ishs.org/icra/index.htm). ICRAs are often managed by horticultural societies such as the Royal Horticultural Society in the United Kingdom, or associations like the American Association of Botanic Gardens and Arboreta (AABGA). They can be responsible for registering cultivars in one, a few or many genera. The AABGA registers Unassigned Woody Ornamentals for over 1000 genera (see Table 6), and the Australian Cultivar Registration Authority registers cultivars for all Australian genera (over 1500). In Table 7 are listed ICRAs that are responsible for registering cultivars for a flora, plant family, plant order or plant habit. Other genera with ICRAs are listed in Table 8.

Table 6: Cultivars of woody plant genera registered by the American Association of Botanic Gardens and Arboreta (excluding those genera with their own ICRA; see Tables 8 and 9)

Abelia	Allanblackia	Apuleia	Ballota
Abroma	Alluaudia	Aquilaria	Balsamocitrus
Abrus	Alniphyllum	Aralia	Banisteriopsis
Abutilon	Alnus	Arbutus	Barringtonia
Acalypha	Aloysia	Arceuthobium	Basutica
Acca	Altingia	Archidendron	Bauhinia
Acer	Amasonia	Arctostaphylos	Beilschmiedia
Acioa	Amblygonocarpus	Arctotis	Bejaria
Acridocarpus	×Amelasorbus	Ardisia	Berberidopsis
Acrocarpus	Amherstia	Argania	Berberis
Acropogon	Amicia	Argyranthemum	Berchemia
Actinodaphne	Ammodendron	Argyrocytisus	Berchemiella
Adansonia	Amomyrtus	Argyrolobium	Bertholletia
Adenanthera	Amoreuxia	Aristolochia	Berzelia
Adenocarpus	Amorpha	Aronia	Beschorneria
Adenostoma	Ampelopsis	Artabotrys	Betula
Adesmia	Amyris	Artemisia	Bignonia
Aegle	Anacardium	Artocarpus	Bixa
Aesculus	Anadenanthera	Asimina	Blepharocalyx
Aetoxicon	Anagyris	Aspidistra	Blighia
Afraegle	Anamirta	Asteromyrtus	Boehmeria
Afzelia	Andrachne	Astiria	Boenninghausenia
Agapetes	Anemopaegma	Atalantia	Bolusanthus
Agarista (Coreopsis)	Angostura	Atraphaxis	Bombax
Agathosma	Aniba	Atriplex	Bosea
Agave	Anisoptera	Atuna	Boswellia
Aglaia	Annona	Aucoumea	Botryostege
Ailanthus	Anogeissus	Aucuba	Bouvardia
Ajania	Anopyxis	Azadirachta	Bowkeria
Akebia	Anthyllis	Azara	Brachyglottis
Alangium	Antiaris	Azorina	Brachystegia
Albizia	Aphanamixis	Baccharis	Brasenia
Alectryon	Aphananthe	Baikiaea	Bravoa
Alhagi	Apodytes	Baillonella	Brexia
Alibertia	Apollonias	Balfourodendron	Brosimum

x = intergeneric hybrid
+ = graft hybrid (intergeneric)

Table 6: Cultivars of woody plant genera registered by the American Association of Botanic Gardens and Arboreta (excluding those genera with their own ICRA; see Tables 8 and 9)

Broussonetia	Capparis	Cercidiphyllum	Citharexylum
Brownea	Caragana	Cercis	Citrofortunella
Bruguiera	Carapa	Cercocarpus	×Citroncirus
Brunfelsia	Cardiandra	Cestrum	Citronella
Brunia	Cardiospermum	Chaenomeles	Citropsis
Brunnichia	Carica	Chamaebatia	Citrus
Bryanthus	Cariniana	Chamaebatiaria	Cladrastis
Bucida	Carmichaelia	Chamaecrista	Clappertonia
Buckleya	Carpenteria	Chamaecytisus	Clausena
Buddleja	Carpinus	Chamaedaphne	Clavija
Bursera	Carpodetus	Chamaelaucium	Clematoclethra
Butea	Carrierea	Chamaespartium	Clerodendrum
Cabomba	Carya	Chiliotrichum	Clethra
Caesalpinia	Caryopteris	Chilopsis	Cleyera
Caldcluvia	Casearia	Chimaphila	Clianthus
Calicotome	Casimiroa	Chimonanthus	Cliftonia
Calliandra	Cassia	Chionanthus	Clitoria
Callicarpa	Cassiope	Chiranthodendron	Clusia
Calligonum	Castanopsis	Chloranthus	Clymenia
Calodendrum	Castanospermum	Chlorocardium	Clytostoma
Caloncoba	Castilla	Chloroxylon	Cneorum
Calophaca	Catalpa	Choisya	Cobaea
Calophyllum	Catha	Chondodendron	Coccoloba
Calycanthus	Cavendishia	Chordospartium	Cocculus
Calycocarpum	Cayratia	Chorisia	Cochlospermum
Calycomis	Ceanothus	Chrysobalanus	Codiaeum
Campomanesia	Cecropia	Chrysolepis	Cola
Campsidium	Cedrela	Chrysophyllum	Coleogyne
Campsis	Cedrelinga	Chrysothamnus	Coleonema
Camptotheca	Ceiba	Chukrasia	Colletia
Campylotropis	Celastrus	Cinchona	Colophospermum
Cananga	Celtis	Cineraria	Colquhonia
Canarium	Cephalanthus	Cinnamomum	Colquhounia
Canella	Ceratonia	Cissus	Colubrina
Cantua	Ceratostigma	Cistus	Colutea

x = intergeneric hybrid
+ = graft hybrid (intergeneric)

Table 6: Cultivars of woody plant genera registered by the American Association of Botanic Gardens and Arboreta (excluding those genera with their own ICRA; see Tables 8 and 9)

Colvillea	Cunonia	Diapensia	Drypis
Combretum	Cupaniopsis	Diatenopteryx	Dulacia
Commiphora	Cussonia	Dichotomanthe	Duranta
Comptonia	Cydista	Dichotomanthes	Durio
Conocarpus	Cydonia	Dichroa	Dysoxylum
Conradina	Cynometra	Dicorynia	Ebenopsis
Copaifera	Cyphanthera	Didieria	Eccremocarpus
Corallospartium	Cyphomandra	Diervilla	Echinospartum
Corchorus	Cyphostemma	Dillenia	Edgeworthia
Cordeauxia	Cyrilla	Dimocarpus	Ehretia
Cordyline	Cytisus	Diosma	Ekebergia
Corema	Dacryodes	Diospyros	Elaeagnus
Coriaria	Dais	Diostea	Elaeocarpus
Cornus	Danae	Dipelta	Eleutherococcus
Corokia	Daniellia	Diplacus	Elingamita
Corylopsis	Daphne	Diploknema	Elliottia
Corylus	Daphniphyllum	Dipterocarpus	Embelia
Corynabutilon	Davidia	Dipteronia	Empetrum
Corynocarpus	Debregeasia	Dirca	Encelia
Cotinus	Decaisnea	Disanthus	Endiandra
Cotoneaster	Decarya	Discaria	Enkianthus
Cotylelobium	Decodon	Distemonanthus	Entada
Couepia	Decumaria	Distictis	Entandrophragma
Couroupita	Deherainia	Distyliopsis	Entelea
Cowania	Delonix	Distylium	Enterolobium
Craibiodendron	Dendromecon	Docynia	Epigaea
+ Crataegomespilus	Dendropanax	Dodecadenia	Ercilla
Crataegus	Dermatobotrys	Dombeya	Erinacea
× Crataemespilus	Derris	Dorstenia	Eriobotrya
Crateva	Desfontainia	Dovyalis	Eriocephalus
Cratoxylum	Desmanthus	Dracophyllum	Eriodictyon
Crescentia	Desmodium	Drapetes	Eriolobus
Crinodendron	Detarium	Drimys	Erythrina
Crossandra	Deutzia	Dryas	Erythrophleum
Cryptocarya	Dialium	Dryobalanops	Escallonia

x = intergeneric hybrid
+ = graft hybrid (intergeneric)

Table 6: Cultivars of woody plant genera registered by the American Association of Botanic Gardens and Arboreta (excluding those genera with their own ICRA; see Tables 8 and 9)

Euclea	Frangula	Gymnocladus	Hydnocarpus
Eucommia	Frankenia	Haematoxylum	Hymenaea
Eugenia	Franklinia	Hagenia	Hypelate
Eumorphia	Fraxinus	Halesia	Hypericum
Euodia	Fremontodendron	Halimiocistus	Hypocalyptus
Euonymus	Freylinia	Halimium	Idesia
Euphorbia	Fumana	Halimodendron	Illicium
Euptelea	Furcraea	Haloxylon	Indigofera
Eurya	Galphimia	Harpephyllum	Inga
Euryops	Garcinia	Harpullia	Intsia
Euscaphis	Gardenia	Harrimanella	Iochroma
Eusideroxylon	Garrya	Harungana	Isoplexis
Eustrephus	Gaultheria	Hedysarum	Itea
Exbucklandia	Gaylussacia	Heimia	Itoa
Exochorda	Geissois	Heisteria	Iva
Fabiana	Geitonoplesium	Helicteres	Jacaranda
Fagraea	Gelsemium	Helipterum	Jacaratia
Fagus	Genista	Helwingia	Jacquinia
Faidherbia	Gesnouinia	Hemiptelea	Jamesia
Falcataria	Glandularia	Heptacodium	Jarilla
Fallugia	Gleditsia	Heritiera	Jasminum
Fatshedera	Globularia	Hermannia	Jateorhiza
Fatsia	Glochidion	Hesperaloe	Juanulloa
Felicia	Glycosmis	Hexachlamys	Kadsura
Fendlera	Gmelina	Hibiscus	Kageneckia
Fendlerella	Gomphostigma	Hildegardia	Kalmiopsis
Filicium	Gonystylus	Hippophae	×Kalmiothamnus
Firmiana	Gordonia	Hoheria	Kalopanax
Flacourtia	Gossweilerodendron	Holboellia	Keckiella
Fontanesia	Grewia	Holodiscus	Kelseya
Forestiera	Greyia	Homalocladium	Kerria
Forsythia	Griselinia	Hopea	Khaya
Fortunearia	Guarea	Hovenia	Kigelia
Fortunella	Guazuma	Hudsonia	Kleinhovia
Fothergilla	Guibourtia	Huodendron	Koelreuteria

x = intergeneric hybrid
+ = graft hybrid (intergeneric)

Table 6: Cultivars of woody plant genera registered by the American Association of Botanic Gardens and Arboreta (excluding those genera with their own ICRA; see Tables 8 and 9)

Kolkwitzia	Litchi	×Malosorbus	Mimusops
Koompassia	Lithocarpus	Malpighia	Mitchella
Lablab	Lithodora	Malvaviscus	Moneses
+Laburnocytisus	Litsea	Mammea	Monodora
Laburnum	Lobostemon	Mangifera	Mora
Lagetta	Loiseleuria	Manilkara	Morella (Myrica)
Lansium	Lophira	Mansoa	Morinda
Lantana	Lophomyrtus	Mansonia	Moringa
Lapageria	Loranthus	Maranthes	Morus
Lardizabala	Loropetalum	Marcetella	Mucuna
Laurelia	Lovoa	(Bencomia)	Mundulea
Laurus	Loxostylis	Marcgravia	Munronia
Lavandula	Luculia	Margyricarpus	Muntingia
Lavatera	Luetkea	Markhamia	Murraya
Lecythis	Luma	Mascagnia	Musanga
×Ledodendron	Lunasia	Maurandella	Mutisia
Ledum	Luzuriaga	Maurandya	Myrceugenia
Leea	Lycium	Maytenus	Myrcia
Leiophyllum	Lyonia	Melia	Myrcianthes
Leitneria	Lyonothamnus	Melianthus	Myrciaria
Lemuropisum	Lysidice	Melicoccus	Myrica
Lepechinia	Lysiloma	Melicope	Myricaria
Lespedeza	Maackia	Melicytus	Myriocarpa
Leucaena	Macfadyena	Meliosma	Myristica
Leucophyllum	Mackaya	Menispermum	Myrrhinium
Leucothoe	Macleania	Menziesia	Myrsine
Leycesteria	Maclura	Meryta	Myrteola
Licania	Macropiper	Mespilus	Myrtus
Ligustrum	Maddenia	Mesua	Nandina
Limonia	Madhuca	Metrosideros	Napoleonaea
Limoniastrum	Maesa	Microberlinia	Nardophyllum
Lindera	Maesopsis	Microcos	Naringi
Lindleya	×Mahoberberis	Microglossa	Neillia
Linnaea	Mahonia	Milicia	Nemopanthus
Liquidambar	Mallotus	Mimosa	Neobalanocarpus

x = intergeneric hybrid
+ = graft hybrid (intergeneric)

Table 6: Cultivars of woody plant genera registered by the American Association of Botanic Gardens and Arboreta (excluding those genera with their own ICRA; see Tables 8 and 9)

Neolitsea	Otoba	Pernettya	Pithecellobium
Nephelium	Ovidia	Perrottetia	Pithecoctenium
Nerium	Oxydendrum	Persea	Planera
Neviusia	Pachira	Pertya	Platanus
Nierembergia	Pachysandra	Petrea	Platonia
Nipponanthemum	Pachystegia	Petrophytum	Platycarya
Nitraria	Paederia	Petteria	Platycrater
Nivenia	Palaquium	Peumus	Pleonotoma
Noltea	Paliurus	Phellodendron	Plinia
Nothofagus	Panax	Philadelphus	Plocama
Notobuxus	Pandanus	× Philageria	Podalyria
Notospartium	Pangium	Philesia	Podranea
Nuxia	Paraserianthes	Phillyrea	Polianthes
Nyssa	Parashorea	Phoebe	Poliomintha
Ochna	Parinari	Photinia	Poliothyrsis
Ochroma	Parkia	Phygelius	Polygala
Ocotea	Parkinsonia	Phylica	Polylepis
Oemleria	Parmentiera	Phylliopsis	Polyscias
Olax	Parrotia	Phyllodoce	Pometia
Oldenburgia	Parrotiopsis	Phyllostylon	Poncirus
Olea	Parthenocissus	× Phyllothamnus	Pourouma
Oncoba	Paullinia	Physocarpus	Pouteria
Ongokea	Paulownia	Picconia	Prasium
Ononis	Paxistima	Picrasma	Prinsepia
Oplopanax	Payena	Pieris	Prioria
Oreopanax	Peltogyne	Pilocarpus	Prosopis
Orixa	Peltophorum	Pimelia	Protium
Orthilia	Pennantia	Pimenta	Prunus
Osmanthus	Pentace	Pinckneya	Pseudobombax
× Osmarea	Pentaclethra	Piper	Pseudocydonia
Osteomeles	Pentactina	Piptadeniastrum	Pseudopanax
Osteospermum	Pentadesma	Piptanthus	Pseudowintera
Ostrya	Pentapetes	Piscidia	Psidium
Ostryopsis	Peperomia	Pisonia	Psoralea
Osyris	Peraphyllum	Pistacia	Ptelea

x = intergeneric hybrid
+ = graft hybrid (intergeneric)

Table 6: Cultivars of woody plant genera registered by the American Association of Botanic Gardens and Arboreta (excluding those genera with their own ICRA; see Tables 8 and 9)

Pteleopsis	Rhamnella	Scaphium	Skimmia
Ptelidium	Rhamnus	Schaefferia	Sloanea
Pterocarpus	Rhaphiolepis	Schefflera	Smilax
Pterocarya	Rhaphithamnus	Schima	Solandra
Pteroceltis	Rhizophora	Schinopsis	Sophora
Pterogyne	Rhodoleia	Schinus	Sorbaria
Pterospermum	Rhodomyrtus	Schisandra	×Sorbaronia
Pterostyrax	Rhodothamnus	Schizolobium	×Sorbocotoneaster
Pterygota	Rhodotypos	Schizophragma	×Sorbopyrus
Ptychopetalum	Rhoicissus	Schleichera	Sorbus
Pueraria	Rhus	Schotia	Sparrmannia
Punica	Ribes	Schrebera	Spartium
Purshia	Rivina	Sclerocarya	Spathodea
Putoria	Robinia	Semecarpus	Sphaeralcea
Putterlickia	Rosmarinus	Semele	Spigelia
Pycnanthus	Rostrinucula	Senna	Spiraea
×Pyracomeles	Rubus	Seriphidium	Spondias
Pyrenaria	Ruscus	Serissa	Stachyurus
Pyrocydonia	Russelia	Serjania	Staehelina
Pyrola	Ruta	Sesbania	Staphylea
+Pyronia	Sageretia	Severinia	Stauntonia
Pyrostegia	Salix	Shepherdia	Stellera
Pyrularia	Salpichroa	Shorea	Stephanandra
Pyrus	Samanea	Shortia	Stephania
Pyxidanthera	Sambucus	Sibiraea	Sterculia
Quassia	Sandoricum	Simaba (Quassia)	Stewartia
Quillaja	Santolina	Simarouba	Stigmaphyllon
Quintinia	Sapindus	Simmondsia	Streptosolen
Quisqualis	Sapium	Sindora	Strongylodon
Randia	Saraca	Sinobambusa	Strychnos
Rapanea	Sarcococca	Sinocalycanthus	Stryphnodendron
Reaumuria	Sarcopoterium	Sinofranchetia	Styrax
Reevesia	Sargentodoxa	Sinojackia	Sutera
Rehderodendron	Sassafras	Sinomenium	Sutherlandia
Retama	Saurauia	Sinowilsonia	Swietenia

x = intergeneric hybrid
+ = graft hybrid (intergeneric)

Table 6: Cultivars of woody plant genera registered by the American Association of Botanic Gardens and Arboreta (excluding those genera with their own ICRA; see Tables 8 and 9)

×*Sycoparrotia*	*Tibouchina*	*Turnera*	*Visnea*
Sycopsis	*Tieghemella*	*Turraea*	*Vitellaria*
Symphonia	*Tilia*	*Turraeanthus*	*Vitex*
Symphoricarpos	*Tipuana*	*Tutcheria*	*Vitis*
Symplocos	*Tococa*	*Ugni*	*Vouacapoua*
Synoum	*Toona*	*Ulex*	*Waltheria*
Synsepalum	*Toxicodendron*	*Ulmus*	*Weigela*
Tabebuia	*Trachelospermum*	*Umbellularia*	*Weinmannia*
Talisia	*Treculia*	*Ungnadia*	*Wigandia*
Tamarindus	*Trema*	*Urera*	*Wikstroemia*
Tamarix	*Trevesia*	*Urostemon*	*Wisteria*
Tapiscia	*Trichilia*	(*Brachyglottis*)	*Withania*
Tecoma	*Trichostigma*	*Vaccinium*	*Xanthoceras*
Tecomanthe	*Tripetaleia*	*Vallea*	*Xanthorhiza*
Tephrosia	*Triphasia*	*Vateria*	*Ximenia*
Terminalia	*Triplochiton*	*Vatica*	*Xylia*
Ternstroemia	*Tripterygium*	*Vella*	*Xylopia*
Tetracentron	*Tristaniopsis*	*Ventilago*	*Xylosma*
Tetradenia	*Tristellateia*	*Verbena*	*Yucca*
Tetradium	*Trochetia*	*Vestia*	*Zanthoxylum*
Tetrapanax	*Trochetiopsis*	*Vinca*	*Zapoteca*
Tetrastigma	*Trochocarpa*	*Virgilia*	*Zelkova*
Theobroma	*Trochodendron*	*Virola*	*Zenobia*
Theophrasta	*Tsusiophyllum*	*Viscum*	*Ziziphus*

x = intergeneric hybrid
+ = graft hybrid (intergeneric)

Table 7: ICRAs that are responsible for registering cultivars for a flora, plant order, plant family, or plant habit

CATEGORY	ICRA
Flora	Australian Cultivar Registration Authority (ACRA) (www.anbg.gov.au/acra/)
Plant order	Conifers including Gingko (www.rhs.org.uk/research/registration_conifers_accepted.asp)
Plant family	• Araceae excluding Calla and Zantedeschia (www.aroid.org/cultivars/index.html) • Bromeliaceae (www.bsi.org) • Gesneriaceae (except Saintpaulia) (www.aggs.org/ir_ges/) • Magnoliaceae (www.magnoliasociety.org/checklist_ndx.html) • Nymphaeaceae (collections2.eeb.uconn.edu/collections/herbarium/reghome.html) • Orchidaceae (www.rhs.org.uk/plants/registration_orchids.asp) • Proteaceae (excluding Australian genera) (www.nda.agric.za/docs/Protea2002/proteaceae_register.htm)
Plant habit	• Bulbous, cormous and tuberous-rooted ornamental plants excluding *Dahlia, Lilium, Narcissus, Nerine* and various Australian genera (www.kavb.nl/index.cfm) • Carnivorous plants (www.carnivorousplants.org/) • Unassigned woody ornamentals (www.bbg.org/sci/taxonomy/registration.html) (see Table 8 for a list of genera)

Table 8: Genera not covered by ICRAs for a flora, plant family, plant order or plant habit

Genus	Contact details
Actinidia	www.hort.cri.nz
Amelanchier	www.ag.usask.ca/departments/plsc/nfdp/index.html
Andromeda	www.heathersociety.org.uk/registration.html
Astilbe	www.holehirdgardens.org.uk
Begonia	www.begonias.org
Bougainvillea	www.iari.res.in
Brugmansia	www.brugmansiaregistry.com/
Buxus	(box) www.boxwoodsociety.org/
Calluna	www.heathersociety.org.uk/registration.html
Camellia	www.camellia-ics.org
Castanea	www.caes.state.ct.us/FactSheetFiles/PlantPathology/fspp063f.htm
Clematis	www.rhs.org.uk/plants/registration_clematis.asp
Clivia	cliviasmith.idx.com.au/
Coprosma	www.rnzih.org.nz/
Curcuma	www.nparks.gov.sg/parks/sbg/par-sbg.shtml
Cyclamen (except Cyclamen persicum)	www.cyclamen.org/indexcs.html
Cyclamen persicum	www.vkc.nl/
Daboecia	www.heathersociety.org.uk/registration.html
Dahlia	www.rhs.org.uk/plants/registration_dahlia.asp
Datura	www.americanbrugmansia-daturasociety.org/
Delphinium	www.rhs.org.uk/plants/registration_delphinium.asp
Dianthus	www.rhs.org.uk/plants/registration_dianthus.asp
× Disberocereus	www.rhs.org.uk/plants/registration_dianthus.asp
Disocactus	www.epiphyllumsociety.org
× Disophyllum	www.epiphyllumsociety.org
× Disoselenicereus	www.epiphyllumsociety.org
Elmerrillia	www.magnoliasociety.org
Epiphyllum	www.epiphyllumsociety.org
Erica	www.heathersociety.org.uk/registration.html
Erodium	www.bpgs.org.uk
Euryale	www.iwgs.org
Eustoma	www.vkc.nl/
Ficus	www.vkc.nl/

Genus	Contact details
Fuchsia	www.americanfuchsiasociety.org/registra.html
Gentiana	www.srgc.org.uk/
Geranium	www.bpgs.org.uk
Gerbera	www.vkc.nl/
Gladiolus	www.gladworld.com
Hamamelis	www.arboretumkalmthout.be/
Hebe	www.rnzih.org.nz/
Hedera (Ivy)	www.ivy.org
Hedychium	www.nparks.gov.sg/parks/sbg/par-sbg.shtml
Heliohebe	www.rnzih.org.nz/
Hemerocallis (daylilies)	www.daylilies.org
Hibiscus rosa-sinensis	www.australianhibiscus.com/Database/register/boot.htm
Hosta	www.hosta.org/About_Hosta/registration_database.htm
Hydrangea	helene.bertrand@inh.fr
Hylocereeae	www.epiphyllumsociety.org
Hylocereus	www.epiphyllumsociety.org
×Hylophyllum	www.epiphyllumsociety.org
Ilex	www.hollysocam.org/registrations.htm
Iris (excluding bulbous species)	www.irises.org
Juglans	www.agriculture.purdue.edu/fnr/HTIRC/Juglans.htm
Kalmia	www.highsteadarboretum.org/
Kmeria	www.magnoliasociety.org
Lagerstroemia	www.usna.usda.gov/Research/Herbarium/Lagerstroemia/Checklist_A.html
Lilium	www.rhs.org.uk/plants/registration_lilium.asp
Liriodendron	www.magnoliasociety.org
Lonicera	blahnik@mze.cz
Magnolia	www.magnoliasociety.org
Malus	www.malus.net
Mangifera indica	www.iari.res.in
Manglietia	www.magnoliasociety.org
Meconopsis	evelyn@thelinns.org.uk
Michelia	www.magnoliasociety.org
Narcissus	www.rhs.org.uk/plants/registerpages/intro.asp
Nelumbo	www.iwgs.org
Nerine	barrieward@aol.com

Genus	Contact details
Nuphar	www.iwgs.org
Nymphaea	www.iwgs.org
Ondinea	www.iwgs.org
Pachylarnax	www.magnoliasociety.org
Paeonia (Peony)	www.americanpeonysociety.org
Passiflora	www.passionflow.co.uk/reg.htm
× Paulheathara	www.epiphyllumsociety.org
Pelargonium	www.begs.org.uk
Penstemon	dlindgren1@unl.edu
Phormium	www.rnzih.org.nz/
Pittosporum	www.rnzih.org.nz/
Plumeria	www.theplumeriasociety.org/dyncat.cfm?catid=2001
Populus	www.fao.org/forestry/index.jsp
Potentilla fruticosa	cdavidson@agr.gc.ca
Prunus mume	chenjymc@public.bta.net.cn
Pseudorhipsalis	www.epiphyllumsociety.org
Pyracantha	www.usna.usda.gov/
Quercus	www.saintmarys.edu/~rjensen/ios.html
Rhododendron	www.rhs.org.uk/plants/registration_rhododendron.asp
Rosa (Roses)	www.ars.org
Saintpaulia (African Violets)	www.avsa.org/
Saxifraga	www.saxifraga.org/
Selenicereus	www.epiphyllumsociety.org
× Seleniphyllum	www.epiphyllumsociety.org
Syringa	www.rbg.ca/pages_sci_conserv/sci_conserv_lregistrar.html
Viburnum	www.usna.usda.gov/
Victoria	www.iwgs.org
Viola (Violets)	www.americanvioletsociety.org
Weberocereus	www.epiphyllumsociety.org

Authors of plant names

Brummitt RK and Powell CE (1992). *Authors of plant names.* Royal Botanic Gardens, Kew.

Botanical and cultivated codes

Brickell CD, Baum BR, Hetterscheid WLA, Leslie AC, Mcneill J, Trehane P, Vrugtman F and Wiersema JH (Eds) (2004). *International Code of Nomenclature for Cultivated Plants*. Bohn, Scheltema & Holkema, Utrecht. (This can be purchased from the International Society for Horticultural Science through the internet at www.ishs.org/sci/icracpco.htm)

McNeill J, Barrie FR, Burdet HM, Demoulin V, Hawksworth DL, Marhold K, Nicolson DH, Prado J, Silva PC, Skog JE, Wiersema JH and Turland NJ (2006). *International Code of Botanical Nomenclature (Vienna Code) adopted by the Seventeenth International Botanical Congress, Vienna, Austria, July 2005*. A.R.G. Gantner Verlag, Ruggell, Liechtenstein. (Accessible on the internet at www.ibot.sav.sk/karolx/kod/0000Viennatitle.htm)

Botanical Latin, pronunciation, name derivations and meanings

Bailey LH (1933). *How plants get their names*. Macmillan, New York. (Although this book is very good, what is said about the Code is dated. It contains a good accented list of names for pronunciation as does the better-known *Manual of Cultivated Plants*. It is now available from Dover Publications, Mineola, NY.)

Baines JA (1981). *Australian plant genera, an etymological dictionary of Australian plant genera*. Society for Growing Australian Plants, Sydney.

Coombes AJ (1985). *Dictionary of plant names*. Collingridge, Twickenham.

Gledhill D (2002) *The names of plants*. 3rd edition. Cambridge University Press, Cambridge.

Hall N and Johnson LAS (1993). *The names of Acacias of New South Wales with a guide to pronunciation of botanical names*. Royal Botanic Gardens, Sydney. (A full account of pronunciation for all botanical names.)

Johnson AT and Smith HA (1984). *Plant names simplified*. Landsmans, Bromyard.

Perrin D (undated, c. 1987). *Dictionary of botanical names. Australian plant names, meaning, derivation and application*. Redcliffe Education Centre, Redcliffe, Queensland.

Radford AE, Dickison WC, Massey JR and Bell CR (1974). *Vascular plant systematics*. Harper & Row: New York.

Sharr FA (1978). *Western Australian plant names and their meanings*. University of Western Australia Press, Nedlands, Western Australia.

Smith AW (1972). *A gardener's dictionary of plant names*. Revised and enlarged by W.T. Stearn. Cassell, London.

Stearn WT (1987). *Botanical Latin*. 3rd edition, revised. David & Charles, Newton Abbot, United Kingdom.

Botanic gardens and herbaria

If you are unsure about a particular name after checking the references listed, contact your local botanic garden or herbarium for advice. A list of herbaria can be found at sciweb.nybg.org/science2/IndexHerbariorum.asp. Botanic garden locations can be found through the Botanic Gardens Conservation International website at www.bgci.org.uk/.

Classification systems

Angiosperm Phylogeny Group (1998). An ordinal classification for the families of flowering plants. *Annals of the Missouri Botanical Garden* **85**, 531–553.

Angiosperm Phylogeny Group (2003). An update of the Angiosperm Phylogeny Group classification for the orders and families of flowering plants: APG II. *Botanical Journal of the Linnean Society* **141**, 399–436.

Brummitt RK (compiler) (1992). *Vascular plant families and genera*. Royal Botanic Gardens, Kew.

Chase MW, Soltis DE Soltis PS, Rudall PJ, Fay MF, Hahn WH, Sullivan S, Joseph J, Molvray M, Kores PJ, Givnish TJ, Sytsma J and Pires JC (2000). Higher-level systematics of the monocotyledons: an assessment of current knowledge and a new classification. In *Monocots – systematics and evolution*. (Eds KL Wilson and DA Morrison) pp. 3–16. CSIRO Publishing, Melbourne.

Cronquist AJ (1981). *An integrated system of classfication of flowering plants*. Columbia University Press, New York.

Cronquist AJ (1988). *The evolution and classification of flowering plants*. 2nd edn. New York Botanical Gardens, New York.

Dahlgren RTM (1983). General aspects of angiosperm evolution and macrosystematics. *Nordic Journal of Botany* 3, 119–149.

Dahlgren RMT, Clifford HT and Yeo PF (1985). *The families of monocotyledons – structure, evolution, and taxonomy.* Springer-Verlag, Berlin.

Kubitzki K (Ed.) (1998). *The families and genera of vascular plants, Volumes I to VI.* Springer-Verlag, Berlin.

Linnaeus C (1753). *Species plantarum.* Laurentii Salvii, Holmiae.

Stevens PF (2001 onwards). *Angiosperm Phylogeny Website.* Version 5, May 2004 (and more or less continuously updated since), www.mobot.org/MOBOT/research/APweb/

Styles BT (Ed.) (1986). *Infraspecific classification of wild and cultivated plants.* Clarendon Press, Oxford.

Takhtajan A (1997). *Diversity and classification of flowering plants.* Columbia University Press, New York.

Plant Breeder's Rights

INTERNATIONAL

The International Union for the Protection of New Varieties of Plants (UPOV), www.upov.int/. (An intergovernmental organisation with headquarters in Geneva, Switzerland.) See *Cultivated Plant Code* for full listing.

ASIA-PACIFIC

Australian Plant Breeder's Rights Office (www.ipaustralia.gov.au/pbr/index.shtml)

Japan: Plant Variety Protection (www.hinsyu.maff.go.jp/english/index.htm)

New Zealand Plant Variety Rights Office (www.pvr.govt.nz)

Plant Varieties Journal, AGPS, GPO Box 84, Canberra, ACT 2601

EUROPE

France: Comité de la protection des obtentions végétales (geves.zarcrom.fr)

Germany: Bundessortenamt (www.bundessortenamt.de/internet20_engl/)

Italy: Ufficio Italiano Brevetti e Marchi (www.minindustria.it)

Netherlands: Raad voor plantenrassen (www.kwekersrecht.nl)

United Kingdom: Plant Varieties Office www.defra.gov.uk/planth/pvs/default.htm

North America
Canadian Plant Breeder's Rights Office (www.inspection.gc.ca/english/
plaveg/pbrpov/pbrpove.shtml#var)
United States Patent and Trademark Office (www.uspto.gov/)
United States Plant Variety Protection Office (www.ams.usda.gov/science/
pvpo/pvpindex.htm)

Trademarks

Asia-Pacific
Intellectural Property Australia (www.ipaustralia.gov.au/)
Intellectual Property Office of New Zealand (www.ipmenu.com/country/
newzealand.htm)

Europe
The Patent Office United Kingdom (www.patent.gov.uk/tm/)

North America
Canadian Intellectual Property Office (strategis.ic.gc.ca/sc_mrksv/cipo/tm/
tm_main-e.html)
United States Patent & Trademark Office (www.uspto.gov/main/
trademarks.htm)

Appendix

Table 9: Examples of different kinds of plant names according to the Codes, including different kinds of plants and where they are growing

The *Botanical Code* governs the genus and specific epithets in Latin form for both wild plants and cultivated plants (except graft chimaeras which are governed by the *Cultivated Plant Code*). The *Cultivated Plant Code* governs additional names in any language other than Latin (except when the Latin name is retained as a cultivar or Group epithet after reassessment of the taxon, or when the cultivar or Group names in Latin form were established before 1 January 1959, or graft chimaeras with Latin names).

Kind of plant	Where the plant is growing	Examples		Comments
		Botanical Code components of name	Cultivated Plant Code components of name	Full name	
Wild plant	In natural habitat	*Gardenia thunbergia*		*Gardenia thunbergia*	A species found naturally in the evergreen forests in South Africa
Wild plant	Cultivated in gardens	*Hardenbergia violacea*		*Hardenbergia violacea*	These plants, although cultivated by humans, are assumed to be genetically unaltered wild plants
		Viburnum tinus		*Viburnum tinus*	
Natural hybrid	In natural habitat usually with populations of both parents present producing hybrid progeny	*Acacia baileyana* × *A. dealbata*		*Acacia baileyana* × *A. dealbata*	A hybrid formula between two species of *Acacia*
		Eucalyptus ×*studleyensis*		*Eucalyptus* ×*studleyensis*	Name with a collective epithet for the hybrid between *Eucalyptus camaldulensis* and *E. ovata*

Natural hybrid	Naturalised in natural habitat, fertile and reproducing without the need for parent populations to be present	*Brachychiton ×carneus*		*Brachychiton ×carneus*	*Brachychiton ×carneus* is a fertile natural hybrid
Cultigen hybrids	Cultivated in gardens	*Camellia ×williamsii*		*Camellia ×williamsii*	A hybrid between the species *Camellia saluenensis* and *C. japonica* originally raised by Mr J.C. Williams in the UK
		Rhododendron laetum × R. aurigerianum	'Wattle Bird'	*Rhododendron laetum* 'Wattle Bird' × *R. aurigerianum*	A deliberate cross between a cultivar and a wild plant
		Magnolia campbellii × M. ×soulangeana	'Alexandrina'	*Magnolia campbellii × M. ×soulangeana* 'Alexandrina'	A deliberate cross between a wild plant and a selected cultivar of a hybrid originally bred by Soulange-Bodin near Paris in 1820
Cultigen hybrids	Naturalised in natural habitat	*Crocosmia ×crocosmiiflora*		*Crocosmia ×crocosmiiflora*	This hybrid, first bred by Lemoine in France in 1880, has escaped from cultivation in some parts of the world including Australia, NZ, south-west Europe and the USA

Kind of plant	Where the plant is growing	Examples			Comments
		Botanical Code components of name	Cultivated Plant Code components of name	Full name	
Ancient cultigens	Cultivated in gardens, agriculture and sometimes naturalised in natural habitat	*Zea mays* *Saccharum officinale*			*Zea mays* (Corn) and *Saccharum officinale* (Sugarcane), ancient cultigens are plants of uncertain history, but are assumed to have been bred and/or selected over a long period. They remain cultigens even if they become naturalised in natural habitats
		Brassica rapa var. *pekinensis* *Brassica rapa*	Pekinensis Group	*Brassica rapa* var. *pekinensis* *Brassica rapa* Pekinensis Group	These are two ways to name Chinese Cabbage. The first reflects how these ancient cultigens were treated solely under the ICBN. The second way, named under both the ICBN and the ICNCP, is recommended for clearly indicating their unclear origins as ancient cultigens
Cultigen Group		*Prunus*	Sato-zakura Group	*Prunus* Sato-zakura Group	The Japanese Flowering Cherries have been long cultivated in Japan, their parentage being unclear but likely to include three or more species. The Group name is written without brackets when referring to the Group as a whole
Cultigen Group: selection		*Prunus*	(Sato-zakura Group) 'Ojochin'	*Prunus* (Sato-zakura Group) 'Ojochin'	The Group name is written with brackets when writing a cultivar name of the Group

Cultivars: selections of wild plants showing a distinct and reproducible part of their natural variation that is considered worthy of a name	Cultivated in a garden	Callistemon pallidus	'Mt Oberon'	Callistemon pallidus 'Mt Oberon'	Callistemon pallidus 'Mt Oberon' is a Lemon Bottlebrush with silvery young growth
		Hardenbergia violacea	'Happy Wanderer'	Hardenbergia violacea 'Happy Wanderer'	Hardenbergia violacea 'Happy Wanderer' is a very vigorous selection of this climber
		Arbutus unedo	'Compacta'	Arbutus unedo 'Compacta'	Arbutus unedo 'Compacta' has a compact form
Cultivars: selections of wild plants showing a distinct and reproducible part of their natural variation that is considered worthy of a name	Naturalised in natural habitat	Populus nigra	'Italica'	Populus nigra 'Italica'	This commonly grown cultivar of Populus nigra has escaped from cultivation into natural habitats of Australia, New Zealand, Canada and the USA

Kind of plant	Where the plant is growing	Examples			Comments
		Botanical Code components of name	Cultivated Plant Code components of name	Full name	
Cultivars: selections of sports from individual plants or of deliberately bred seedlings of a species	Cultivated in a garden	*Camellia japonica*	'Great Eastern'	*Camellia japonica* 'Great Eastern'	An Australian *Camellia* cultivar first listed in a nursery catalogue in 1872
Cultivars: selections of cultigen hybrids	Cultivated in gardens	*Abelia* ×*grandiflora*	'Francis Mason'	*Abelia* ×*grandiflora* 'Francis Mason'	A selection with yellow and green leaves from the cross *Abelia chinensis* ×*A.uniflora* raised in New Zealand
		Anigozanthos	'Bush Emerald'	*Anigozanthos* 'Bush Emerald'	Selection from the cross *Anigozanthos viridis* ×*A.manglesii* by Merve Turner, Victoria, Australia
		Malus ×*domestica*	'Jonathon'	*Malus* ×*domestica* 'Jonathon' Apple 'Jonathon'	Common names can be used when there is no possibility of confusion

Graft chimaeras: plants consisting of tissue with more than one genetic origin	Cultivated in gardens	*Crataegus + Mespilus +Crataegomespilus*	*Crataegus + Mespilus +Crataegomespilus*	Two ways of naming a graft chimaera with tissue from the genera *Crataegus* and *Mespilus*
Graft chimaeras: cultivar selection	Cultivated in gardens	*+Laburnocytisus* 'Adamii'	*+Laburnocytisus* 'Adamii'	A named selection of the graft chimaera between *Laburnum* and *Cytisus*

Glossary

This is a list of specialist terms used in this publication together with others that are frequently used in relation to plant nomenclature. The list is based on the glossary of the 2004 edition of the *Cultivated Plant Code*, where a more complete glossary is given.

addition sign
in nomenclature, the symbol used to indicate a graft-chimaera.

artificial classification
one that is proposed solely for ease of use and which does not profess to demonstrate natural relationships.

artificial selection
selection by humans, often after breeding, of plants with particular desired characteristics.

author
the person to whom a name or publication is attributed.

author abbreviation
an abbreviation of the name of an author used in a citation; there are now standardised author abbreviations (see Brummitt and Powell 1992).

author citation
a statement of the name(s) of the author(s) responsible for the publication or the establishment of a name.

back-cross
the cross of a hybrid with one of its own parents.

binomial
the scientific name of a species consisting of two words, the first being the name of the genus to which a species belongs and the second being the epithet given to that species to distinguish it from others in the same genus.

Botanical Code
International Code of Botanical Nomenclature, the book containing the international set of rules that provides for the formation and use of the scientific names in Latin of organisms treated as plants.

category
(in cultigen nomenclature) a division (rank) in a system of cultigen classification.

chimaera
an individual composed of two or more genetically different tissues in close association.

circumscribe
make a circumscription.

circumscription
a statement of the diagnostic limits of a taxon.

classification
the systematic grouping of items; a system in which items may be grouped.

clone
two or more individuals, originally derived from one plant by asexual propagation, and which remain genetically identical.

Code
one of the international codes of nomenclature, generally referring to its most recent edition.

code-name
an epithet without evident meaning that is made up of a sequence of connected letters and/or numbers.

collective name
the single designation covering the progeny of a particular hybrid, cf. grex.

colloquial name
a name that is used locally but not widely enough to be recognised in the general dictionaries of the language concerned, cf. vernacular name, common name.

combination
the name of a taxonomic unit below the rank of genus, consisting of the name of the genus with one or two epithets.

commercial synonym
a marketing name chosen as an alternative to the scientific name, in Australia generally used in relation to Plant Breeder's Rights. See also trade designation.

Commission for the Nomenclature of Cultivated Plants
a body promoted by the International Union of Biological Sciences that formulates the *International Code of Nomenclature for Cultivated Plants.*

Commission for Nomenclature and Cultivar Registration
a body of the International Society for Horticultural Science that deals with matters connected with plant naming and promotes the registration of names of cultivated plants.

common name
the name widely used in any language in place of a scientific name and which is generally found in non-technical dictionaries of that language, cf. colloquial name, vernacular name.

conserved name
a name that, although otherwise contrary to the rules of a *Code*, must be adopted as being the accepted name, by the ruling of a body responsible for such decisions.

cross
the act of hybridisation (verb); a hybrid (noun).

cultigen
a collective term for all plants that have been deliberately altered in some way by humans and therefore have one or more characteristics that are different from those of their wild relatives; human-altered plants.

cultigen name
The full name of a cultigen including the Latin component governed by the *Botanical Code* together with the cultivar and/or Group epithets governed by the *Cultivated Plant Code*.

cultivar
a plant with distinct and desirable characteristics that can be reproduced reliably and maintained in cultivation. Defined by the *Cultivated Plant Code* as a rank – 'the primary category of cultivated plants whose nomenclature is governed by the *Cultivated Plant Code*' – and as a taxon – 'an assemblage of plants that has been selected for a particular attribute or combination of attributes and that is clearly distinct, uniform, and stable in these characteristics, and that, when propagated by appropriate means, retains those characteristics'.

cultivar epithet
the defining part of a name that denotes a cultivar.

Cultivated Plant Code
International Code of Nomenclature for Cultivated Plants, the book containing the international set of rules that provides for the formation and use of the scientific names of cultigens.

culton
a taxon of cultigens. Defined in the *Cultivated Plant Code* as 'a systematic group of cultivated plants which is based on one or more user criteria: a word parallel to "taxon" but used solely for a taxonomic unit whose nomenclature is governed by the *Cultivated Plant Code*'.

date of a name
the date of establishment of a cultivar, graft-chimaera or Group.

date of publication
the date on which printed matter became available to the general public or to botanical libraries.

denomination class
the taxonomic unit in which cultivar and Group epithets may not be duplicated except in special circumstances.

description
a statement of the characters of a particular taxon; an expanded diagnosis.

descriptor
a word or phrase appended to the name of a taxon which is used to separate an element such as flower colour.

determine
to perform an identification.

determination
an identification.

diagnosis
a statement which, in the opinion of its author, distinguishes one taxon from another.

diagnostic characters
the features that distinguish a particular taxon from others.

DUS test
the criteria of distinctness, uniformity and stability by which a new cultivar is examined for statutory purposes such as the granting of Plant Breeder's Rights.

epithet
the final word or combination of words in a name that denotes an individual taxon, cf. cultivar epithet and species epithet.

established name
a name that meets the requirements of the *Cultivated Plant Code*.

establishment
a prime principle of cultigen nomenclature whereby, on publication, certain criteria must have been fulfilled before considerations of acceptability are allowed.

exotic
referring to a plant that is not native to a particular region.

genetically modified organism
an organism with new characters following he implantation of alien genetic material.

germplasm
hereditary material transferred to the offspring via the gametes.

graft-chimaera
a plant consisting of tissue from two or more different taxa in close association, produced by grafting.

grafting
the (usually deliberate) fusion of tissue from two or more different plants.

grex
a type of Group used in orchid nomenclature applied to the progeny of an artificial cross from specified parents. (pl. greges)

Group
a formal category denoting an assemblage of cultivars, individual plants, or assemblages of plants on the basis of defined similarity, cf. grex.

herbarium
a collection of botanical specimens; the housing for such specimens.

herbarium specimen
a (usually dried) botanical specimen kept in a herbarium.

hierarchy
a series of progressively more inclusive ranks.

homonym
one of two or more names or epithets spelled, or deemed to be spelled, exactly like another name or epithet, but which is used for a different taxon of the same rank.

hybrid
the result of a cross between differing plants or taxa.

hybrid formula
the names of the parent taxonomic units of a hybrid linked with a multiplication sign (or cross).

illegitimate name
a name to be rejected for not fulfilling the requirements of the *Botanical Code*.

infraspecific
pertaining to any taxon below the rank of species.

International Code of Botanical Nomenclature (Botanical Code)
the international set of rules that provides for the formation and use of the scientific names, in Latin, of organisms treated as plants.

International Code of Nomenclature for Cultivated Plants (Cultivated Plant Code)
the international set of rules that provides for the formation and use of the scientific names of cultigens.

international registrar
the person appointed by an International Cultivar Registration Authority to carry out its registrations.

International Cultivar Registration Authority (ICRA)
an organisation appointed by the International Society for Horticultural Science (ISHS) Commission for Nomenclature and Cultivar Registration to be responsible for registering cultivar and Group names within defined taxa.

International Society for Horticultural Science
the organisation (a scientific member of the International Union of Biological Sciences) established to promote the science of horticulture.

introducer
of a cultivar: the person or organisation who first distributes a cultivar.

legitimate name
applied in the *Botanical Code* to names that are in accordance with the rules of nomenclature, i.e. ones that are not defined as illegitimate.

line
a plant breeding term used to describe plants resulting from repeated self-fertilisation or inbreeding.

lumping
to treat as members of a single taxon elements which have been previously considered as belonging to more than one taxon.

maintenance
sometimes used for a seed-raised cultivar which, although not differing from an existing cultivar, requires a name.

misapplied name
a plant name which has been incorrectly applied; a name that has been perpetuated in a sense not intended by its original author, cf. misidentification.

misidentification
an incorrect determination of a plant name.

modern language
one currently in use.

multiline
a plant breeding term used to describe a cultivar that is made up of several closely related lines.

multiplication sign
in nomenclature, the symbol used to indicate a hybrid.

nomenclatural hierarchy
the consecutively more inclusive ranks of taxa defined by some *Codes* of nomenclature.

nomenclatural standard
a specimen or other item to which the name of a cultivar or Group is permanently attached.

nomenclatural type
under the *Botanical Code* that element to which the name of a taxon is permanently attached, whether as a correct name or as a synonym, and which fixes the application of a name. The nomenclatural type is not necessarily the most typical or representative element of a taxon.

Plant Breeder's Rights (Plant Variety Rights)
a breeder's legal protection over the propagation of a cultivar, abbreviated to PBR.

plant patent
a grant of right, available in some countries, which provides a means of control over a new plant's propagation and sale for a given period.

printed matter
text or illustrations mechanically reproduced by printing in quantity and in intentionally permanent form.

priority
a prime principle of nomenclature whereby the earliest established name takes precedence over later names for the taxon at a particular rank.

provisional name
one that is not established but is proposed in anticipation of the recognition of a taxon with a particular circumscription, position, or rank.

publication
a principle of nomenclature in a *Code* where certain rules must be fulfilled before establishment is assessed, usually achieved by the distribution of

dated printed matter available to the community; the act of distributing printed matter into the public domain.

publish
to issue a publication; to place names and other nomenclatural matter in the public domain.

published
of a name, one that fulfils the requirements of publication.

rank
a level within the nomenclatural hierarchy.

Recommendations
in the *Cultivated Plant Code* and *Botanical Code*, regulations that are encouraged.

registered trademark
a trademark that has been formally accepted by a statutory trademark authority.

registration
the act of recording a new name or epithet with a registration authority.

rejected name
one that is not to be used as a result of failure to comply with certain Rules.

rootstock
the living material on which a scion is grafted, cf. scion.

Rules
in a *Code* the regulations which must be followed.

scientific name
the name of a scientific unit formed and maintained under the rules of the *Codes*.

scion
the shoot, bearing buds, that is used in grafting.

selection
a plant or assemblage of plants that has been isolated on the basis of one or more desirable characteristic(s).

sensu lato
in a broad sense.

sensu stricto
in a narrow sense.

species
the basic category in the nomenclatural hierarchy of wild plants.

species name
the name of the genus (e.g. *Banksia*) together with the specific epithet (e.g. *repens*) is the species name (e.g. *Banksia repens*).

specimen
a plant, or part of a plant, preserved for scientific study.

sport
an apparent mutation that has ocurred on part of a plant.

Standard Portfolio
a repository, usually a folder, in which the nomenclatural standard and its associated information are kept together.

strain
generally refers to seed-raised cultivars that can hardly, if at all, be distinguished from existing cultivars.

synonym
an established (validly published) name denoting a taxonomic unit in a given taxonomic position that is not the accepted (correct) name. In horticulture it can also refer to a name that has been widely, possibly incorrectly, applied to a plant; outdated or 'alternative' names.

synonymy
a list of synonyms.

taxon
the international abbreviation for the words 'taxonomic unit' (pl. taxa).

taxonomic unit
a group into which a number of similar individuals may be classified.

trade designation
a device that is used to market a plant when the original name is considered unsuitable for marketing purposes.

trademark
a letter, number, word, phrase, sound, smell, shape, logo, picture, aspect of packaging or combination of these used to distinguish the goods or services of one trader from those of others.

transcription
to copy precisely from one written work to another; the rendering in written form of sounds of human speech, especially of languages emplying ideographic or phonetic characters.

translation
changing the words of one language into those of another language.

transliteration
changing the words of one alphabetical script into another alphabetical script letter by letter.

typification
the act of designating or selecting a nomenclatural type.

type, see nomenclatural type

undetermined
of a specimen, not identified.

UPOV
acronym for the Union Internationale pour la Protection des Obtentions Végétales (the International Union for the Protection of New Varieties of Plants), the international organisation charged with overseeing the administration of Plant Breeder's Rights.

variant
a plant that shows some measure of difference from the characteristics associated with a particular taxonomic unit.

varietas
the category in the nomenclatural hierarchy between species (*species*) and form (*forma*).

variety
term used in some national and international legislation to denominate a clearly distinguishable taxonomic unit below the rank of species; generally, in this context, a term equivalent to cultivar. When used in a strictly botanical (non-legislative) sense, see *varietas*. Sometimes used in a very loose sense to refer to any kind of plant.

vernacular name
generally understood as a common name but defined by the *Cultivated Plant Code* as a name derived from the translation of the scientific name into a local language.

voucher specimen
a nominated specimen representing the plant or taxonomic unit mentioned in the text.

wild plant
a plant growing naturally in the wild, or occasionally brought into cultivation but unaltered by deliberate human activity, cf. cultigen.

witches broom
a mass of congested, often stunted, stems and foliage on a plant, caused by genetic malformation in the growing shoots.

References

Bailey LH (1924). *Manual of cultivated plants most commonly grown in the continental United States and Canada*. Macmillan, New York.

Bayley I. West Indian Weed Song. Quoted in Morton JF (1981). *Atlas of Medicinal Plants of Middle America*. Charles C. Thomas, Springfield, Illinois.

Brickell CD, Baum BR, Hetterscheid WLA, Leslie AC, McNeill J, Trehane P, Vrugtman F and Wiersema JH (Eds) (2004). *International code of nomenclature for cultivated plants*. Bohn, Scheltema & Holkema, Utrecht.

Brummitt RK and Powell CE (Eds) (1992). *Authors of plant names*. Royal Botanic Gardens, Kew.

Cronquist A (1981). *An integrated system of classification of flowering plants*. Columbia University Press, New York.

Fletcher H, Gilmour JS, Lawrence GHM, Little Jr EL, Nilsson-Leissner G and de Vilmorin R (1958). *International Code of Nomenclature for Cultivated Plants formulated and adopted by the International Commission for the Nomenclature of Cultivated Plants of the International Union of Biological Sciences*. Regnum Vegetabile 10.

Flora of Australia Editorial Committee (1981–). *Flora of Australia*. CSIRO Publishing, Melbourne.

Flora of North America Editorial Committee (1993–). *Flora of North America north of Mexico*. Oxford University Press, New York.

Geddie W. (Ed) (1959). *Chambers 20th Century Dictionary*. W & R Chambers, Edinburgh.

Greuter W, Barrie FR, Burdet HM, Chaloner WG, Demoulin V,
Hawksworth PM, Jørgensen PM, Nicolson DH, Silva, PC, Trehane P
and McNeill J (Eds) (1994). *International code of botanical nomenclature
(Tokyo code) adopted by the Fifteenth International Botanical Congress,
Yokohama, August–September 1993*. Koeltz Scientific Books, Königstein,
Germany.

Greuter W, McNeill J, Barrie FR, Burdet WG, Demoulin V, Filgueiras TS,
Nicolson DH, Silva PC, Skog JE, Trehane P, Turland NJ and
Hawksworth DL (Eds) (2000). *International code of botanical
nomenclature (St Louis code) adopted by the Sixteenth International
Botanical Congress, St Louis, Missouri, July–August 1999*. Koeltz
Scientific Books, Königstein, Germany.

Hetterscheid WLA and Brandenburg WA (1994). The culton concept:
setting the stage for an unambiguous taxonomy of cultivated plants.
Acta Horticulturae **413**, 29–34.

Lord T, Armitage J, Cubey J, Grant M, Whitehouse C (Eds) (2004). *RHS
Plant Finder 2004–2005*. 17th edn. Dorling Kindersley, Melbourne.

Lumley PF and Spencer RD (1990). *Plant names: a guide to botanical
nomenclature*. Royal Botanic Gardens, Department of Conservation
and Environment, Melbourne.

Lumley PF and Spencer RD (1991). *Plant names: a guide to botanical
nomenclature*. 2nd edn. Royal Botanic Gardens, Department of
Conservation and Environment, Melbourne.

McNeill J, Barrie FR, Burdet HM, Demoulin V, Hawksworth DL, Marhold
K, Nicolson DH, Prado J, Silva PC, Skog JE, Wiersema JH and Turland
NJ (Eds) (2006). *International code of botanical nomenclature (Vienna
Code) adopted by the Seventeenth International Botanical Congress, Vienna,
Austria, July 2005*. A.R.G. Gantner Verlag, Ruggell, Liechtenstein.

Stearn WT (1953). *International Code of Nomenclature for Cultivated Plants
formulated and adopted by the International Botanical Congress
Committee for the Nomenclature of Cultivated Plants and the
International Committee on Horticultural Nomenclature and Registration
at the Thirteenth International Horticultural Congress, London, September
1952*. Royal Horticultural Society, London.

Stearn WT (1992). *Botanical Latin*. 4th edn rev. David & Charles, London.

Stearn WT (1986). Historical survey of the naming of cultivated plants.
Acta Horticulturae **182**, 18–28.

Trehane P, Brickell CD, Baum BR, Hetterscheid WLA, Leslie AC, McNeill
J, Spongberg SA and Vrugtman F (1995). *International code of

nomenclature for cultivated plants. Regnum Vegetabile 133. Quarterjack Publishing, Wimborne, UK.

Tutin TG, Heywood VH, Burges NA, Valentine DH, Walters SM and Webb DA (Eds) (1964a). *Flora Europaea. Volume 1 Lycopoiaceae to Platanaceae.* Cambridge University Press, Cambridge.

Tutin TG, Burges NA, Chater AO, Edmondson JR, Heywood VH, Moore DM, Valentine DH, Walters SM and Webb DA (Eds) (1964b). *Flora Europaea. Volume 1 Psilotaceae to Platanaceae. Second edition.* Cambridge University Press, Cambridge.

Tutin TG, Heywood VH, Burges NA, Moore DM, Valentine DH, Walters SM and Webb DA (Eds) (1968). *Flora Europaea. Volume 2 Rosaceae to Umbelliferae.* Cambridge University Press, Cambridge.

Tutin TG, Heywood VH, Burges NA, Moore DM, Valentine DH, Walters SM and Webb DA (Eds) (1972). *Flora Europaea. Volume 3 Diapensiaceae to Myoporaceae.* Cambridge University Press, Cambridge.

Tutin TG, Heywood VH, Burges NA, Moore DM, Valentine DH, Walters SM and Webb DA (Eds) (1976). *Flora Europaea. Volume 4 Plantaginaceae to Compositae (and Rubiaceae).* Cambridge University Press, Cambridge.

Tutin TG, Heywood VH, Burges NA, Moore DM, Valentine DH, Walters SM and Webb DA (Eds) (1980). *Flora Europaea. Volume 5 Alismataceae to Orchidaceae (Monocotyledones).* Cambridge University Press, Cambridge.

Index